TOWARD THE HABIT OF TRUTH

The Commonwealth Fund Book Program
gratefully acknowledges
the assistance of The Rockefeller University
in the administration of the Program.

Mahlon Hoagland

✣ ✣ ✣ ✣ ✣ ✣ ✣ ✣ ✣ ✣ ✣ ✣ ✣ ✣ ✣

TOWARD THE HABIT OF TRUTH

A Life in Science

✣ ✣ ✣ ✣ ✣ ✣ ✣ ✣ ✣ ✣ ✣ ✣ ✣ ✣ ✣

A volume of
A COMMONWEALTH FUND
BOOK PROGRAM
under the editorship of Lewis Thomas, M.D.

W·W·Norton & Company

NEW YORK LONDON

Printed in the United States of America.

The text of this book is composed in Caledonia,
with display type set in Baskerville.

Composition and manufacturing by
The Maple-Vail Book Manufacturing Group.

Book design by Tere LoPrete.

FIRST EDITION

Library of Congress Cataloging-in-Publication Data

Hoagland, Mahlon B.
Toward the habit of truth : a life in science / Mahlon Hoagland. —
1st ed.
p. cm. — (The Commonwealth Fund Book Program)
Bibliography: p.
Includes index.
1. Hoagland, Mahlon B. 2. Biochemists—United States—Biography.
3. Molecular biologists—United States—Biography. I. Title.
II. Series: Commonwealth Fund Book Program (Series)
QP511.8.H63A3 1990
574.19'2'092—dc20
[B] 89-9353

ISBN 0-393-02754-6

W. W. Norton & Company, Inc., 500 Fifth Avenue, New York, N. Y. 10110

W. W. Norton & Company Ltd., 37 Great Russell Street, London WC1B 3NU

1 2 3 4 5 6 7 8 9 0

For Tess

Contents

Acknowledgments

I am grateful to the Commonwealth Fund for its support in the writing of this book and to the following colleagues, not named in my story, who helped me and gladdened me along the way: Brigitte Askonas, Anita Aspen, Penny Bishop, Lucy Comly, Gibbons Cornwell, Linda Dolan, Marc Dresden, Jacqueline Foss, John Gardner, Lisalotte Hecht, Helene Hill, Margaret Hood, Marion Horton, Arnold Meisler, Robert Nolan, Lorraine Pfefferkorn, Jesse Scott, Ruth Segal, Harold Sox, Masamichi Takagi, and Burton Tropp. I thank Caroline Lupfer for her able editorial assistance and Susan Ricker for her skillful and patient typing of the manuscript.

Introduction

The emergence of medicine from an empirical set of unsubstantiated doctrines and dogmas into the company of proper scientific disciplines had its beginnings toward the end of the nineteenth century, but genuine progress toward a comprehensive understanding of the underlying mechanisms of human disease has become a discernible reality only within the past fifty or so years. While still incomplete, needing all sorts of new and unpredictable information for assurance of its future, the prospects for that future seem brighter today than at any earlier period in the profession's long history. Virtually every advance that has been made thus far has resulted from the new insights brought in from the other basic biomedical sciences. We are now partway along in a full-scale revolution in biology, and the practice of medicine is already being transformed, almost from day to day, by the cascades of new theories trailed by new facts.

The Commonwealth Fund decided, several years back, to sponsor a series of books written by working scientists whose personal participation in the scientific upheaval of this century had made a difference. The author of this volume is such a scientist, and his story is an important one. Dr. Mahlon Hoagland's career provides him with the perspective of a laboratory bench scientist, a physician, and an administrator of a complex and successful research institution with its own engrossing history.

The advisory committee for the Commonwealth Fund Book Program, which recommended the sponsorship of this volume, consists of the following members: Alexander G. Bearn, M.D.; Donald S. Fredrickson, M.D.; Lynn Margulis, Ph.D.; Maclyn McCarty, M.D.; Lady Jean Medawar; Berton Roueché; Frederick Seitz, Ph.D.; and Otto Westphal, M.D. The publisher is represented by Edwin Barber, director of the Trade Department at W. W. Norton & Company, Inc. Suzanne H. Heyd serves as administrative assistant. Margaret E. Mahoney, president of the Commonwealth Fund, has actively supported the work of the advisory committee at every turn.

Lewis Thomas, M.D., Director

Alexander G. Bearn, M.D., Deputy Director

The habit of testing and correcting the concept by its consequences in experience has been the spring within the movement of our civilization. . . . In science and in art and in self-knowledge, we explore and move constantly by turning to the world of sense to ask, Is this so? This is the habit of truth, always minute yet always urgent, which for four hundred years has entered every action of ours; and has made our society and the value it sets on man, as surely as it has made the linotype machine and the scout knife and King Lear and the *Origin of Species. . . .*

—JACOB BRONOWSKI,
Science and Human Values

Preface

The opportunity given me by the Commonwealth Fund to convey to a wider audience something of my experience as a scientist seemed at first to be a pleasant enough task. It should not, after all, be difficult to write about oneself. But the process has proved more trying than expected. First, I was appalled at how poorly my memory served me. Many of the more personal aspects of my relationship to my profession seemed to have gotten buried in the urgency of the astonishing onrush of science during my lifetime. Second, the "habit of truth" can be sorely tested when one's own motivations and behavior are analyzed. What were the *real* forces that moved me, a not very scholarly youth, toward science—and later kept me at it? How to weigh the influence of my scientist father, whose role in my professional life I had spent much time vigorously denying?

How tempting it also was in hindsight to make achievement appear to be the consequence of a series of prescient steps—the way we write about them in scientific papers—when in truth there had been much fumbling in the dark. In addition, there was the troublesome fact that once a major discovery has been made and recorded, the emotion, the action, and the surprises fade into the dry pages of textbooks. How does one bring back the excitement and wonder for readers today, bearing in mind the often unresolvable conflict between the desire to make the experience comprehensible to nonscientists and the need to provide sufficient detail to convey the substance of discovery?

Having produced what I judged to be a reasonable account of my life in science, I stepped back and was struck by the need for a context. So I give one here in the form of some general reflections on the nature of the scientific milieu and process.

The late distinguished biologist Sir Peter Medawar has written that what nonscientists *think* scientists do and what many learned academics *say* scientists do, and indeed what scientists themselves say they do, cannot be relied upon. To understand what science is really about, he suggests, we must eavesdrop. Thus, in any laboratory we might hear the following exchange:

"What gave you the idea of trying ———?"

"What happens if you assume that ———?"

"Actually your results can be accounted for on a quite different hypothesis."

"Obviously a great deal more work has got to be done before ———."

"I don't seem to be getting anywhere."

This kind of talk does not suggest that scientists hunt for facts or follow prescribed methods, still less that they are busy formulating laws. They are instead "building explanatory structures, *telling stories* which are scrupulously tested to see if they are stories about real life."[1]

A scientist quickly realizes that there is no single acceptable way of getting at truth. Each explorer lays out his own itinerary into the unknown: identifies a problem, produces ideas—imagined scenarios—of the way things might actually be, and then tests the ideas by experiment. One key to success is the selection of a good system, or model, that can be manipulated and controlled to give clear answers to incisive questions. As François Jacob has written, in science "one endlessly play[s] at setting up a fragment of the universe which the experiment . . . rudely correct[s]."[2] The process is a wholly natural, if more sophisticated and arduous, extension of the one we all use from birth, intuitively, to build our picture of reality. The process is *active:* in Jacob Bronowski's words, "Science does not watch the world, it tackles it."[3]

The questions scientists ask generally take the form of predictions: if I do this, that should happen. If a prediction is borne out by experiment, it builds confidence in the idea—but does not necessarily prove it right. The failure to affirm a prediction by experiment suggests there is something seriously wrong with the idea that generated it. Thus, good ideas often are those that suggest ways they can be proved false. If a hypothesis offers *no* way to prove itself true or false, it is not useful scientifically. It should be thrown away; it is a belief. Knowledge advances on the wings of testable ideas, not of beliefs.

The conclusions scientists arrive at are statements that have a greater or lesser probability of reflecting reality; they are never certainties. They gain strength as accumulating tests continue to verify them or prove reasonable alternatives to be false. We see not what nature actually is but the extent to which it agrees with our ideas. Thus, the possibility that some of our most cherished truths may someday turn out to be wrong can never be completely ruled out.

While scientists strive for objectivity, they are subject to the same emotions as in other endeavors: frustration, disappointment, exasperation, astonishment, delight—and lots of wishful thinking. After all, the unknown vastly exceeds the known; we start with few and often deceptive clues, and proof is devilishly hard to come by. Science inevitably is a generator of surprises. The likelihood of selecting a really "hot" problem ripe for the solving is not high, and the chances of having the right ideas for solving it are no better. For every clean, aesthetically satisfying, prediction-verifying experiment, there will be a hundred duds that require a return to the drawing board. Much of the time things go just plain wrong: ideas are not incisive enough, the apparatus breaks down, the wrong ingredients are added, the test tube cracks, and so on. Failed experiments have value, of course. They are signposts telling others where not to go; and they are a kind of personal album of the excitement, hopes, and optimism that drive the whole process.

Science obviously places a high premium on honesty. Misconduct in science in the form of deliberately faked experiments is rare. In the form of sloppiness, carelessness, bias, and the influence of wishful thinking, however, it is not so rare. As "big science" burgeons with large teams of workers, intense competitiveness, and a growing influence of the profit motive, scientists will have to watch their behavior more carefully.

There is an even more subtle form of dishonesty in science—a kind of distortion of history. Often we write up our work in such a way as to put our performance in the most favorable light; there seems to be a need to make our story appear as prescient as possible. A few scientists have, however, made the effort to "tell it like it is"—with salutary effect. A pioneer in the practice is James D. Watson. His enormously successful book *The Double Helix* (which he originally entitled *Honest Jim*) is an exceptionally candid, no-holds-barred personal account of his journey with Francis Crick to the revelation of the structure of DNA.[4]

Scientists usually take for granted the spirit of optimism that attends modern science, a spirit firmly rooted in the phenomenal success of the enterprise. I find myself often wondering, with admiration and awe, how my scientific brethren of earlier times—hemmed in as they were by superstition, dogma, weird nontestable explanations, and the hostility of established authority and lacking today's easy access to companionship in the search—summoned the extraordinary courage to ask nature straightforward questions.

The widespread cultural faith that nature *is* comprehensible, that the unknown *is* knowable, is relatively new in human history, going back not much more than three hundred years, to Isaac Newton. Consider the terror that nature in the raw must have caused our ancestors and how fiercely over the centuries free inquiry has been feared and resisted by religious authority and popular superstition. I marvel at the audacity of the lone scientist searching for answers under the eye of an omnipotent God reputed to be easily able to keep the truth from anyone.

Sensitivity to one's own ignorance is an essential state of mind for the scientist, and highly recommended for the nonscientist as well. We honor those explorers of the past when we practice healthy skepticism. We are not likely to ask serious questions or make meaningful predictions without first assessing what we do not know. Belief is the easy way; it "explains" everything by explaining nothing. My colleague Rollin Hotchkiss remembers as a boy bringing a clock to his father and asking him why it had stopped. His father removed the back, and a fly fell out on the table. Said his father, "Well, of course, the engineer's dead!" The dead engineer is still with us in the facile nonexplanations used by such pseudosciences as creationism, astrology, occult and so-called psi phenomena, scientism, and so on and on. Still, we have come a long way.

The pleasure in the exercise of skill to transform idea to substance is common to art and science. My lifetime habit of doing sculpture alongside science has convinced me that the common enjoyment is indeed in the action: manipulating material and technique to uncover simplicity, order, and harmony below the surface of things. In sculpture, a form of one's imagining emerges from the density and obscurity of a material, just as in science a new vision of reality arises from mystery by the action of experimentation.

It is easy to oversentimentalize the similarities of art and science. In science, an idea can become substance only if it fits into a dynamic accumulating body of knowledge—a progression of understanding. Each new piece of work is subject to validation in respect to its compatibility with the bigger picture. It is inspected. tested, tentatively accepted, modified, perhaps discarded. There is really no equivalent progress or cumulativeness in art. Art may progress in technique or indebtedness to predecessor but not in content. In art, creator and created content are inseparable.

In science, however, the discovery is uniquely the discoverer's only in terms of priority and in the way it was made. Its content could have—would have—been found by others. If Columbus had not made the trip, someone else would have; indeed, he was not the first anyway. If Newton, Darwin, and Einstein had not been around, others would have come up with insights that in the end would have built the same edifice of knowledge. Only the history of the construction process would be different. Furthermore, science is rich in examples of simultaneous, independent discoveries, such as Darwin's and Wallace's independently attained visions of natural selection.

Gunther Stent has argued that scientific discoveries are as uniquely a scientist's as works of art are an artist's. He quotes Medawar, for example: "The great thing about [Watson and Crick's] discovery [of the structure of DNA] was its completeness, its air of finality. If Watson and Crick had been seen groping toward an answer, . . . if the solution had come out piecemeal instead of in a blaze of understanding: then it would still have been a great episode in biological history, but something more in the common run of things; something splendidly well done, but not done in the grand romantic manner" that it, in fact, was.[5] Stent also holds that two discoveries arrived at simultaneously are rarely, if ever, identical and therefore are uniquely each discoverer's.

It seems to me that here Stent confuses content with manner of discovery. Of course, every scientific discovery is unique in the *way* it is made. Medawar's comment supports the view that there is little unique about Watson and Crick's relation to their contribution beyond priority and the style in which it was made—which was unquestionably magnificent. And as far as simultaneous discoveries are concerned, their similarities are far more significant than their differences. Wallace may not have had the historical impact that Darwin did, but he independently came up with essentially the same germinal idea—the same content—that provided all of biology with a unifying theme.

In the march of scientific discovery, then, its artisans blend into history like the builders of the great cathedrals. Scientists must be pretty high on egotism to avoid acknowledging their own expendability. This reality, together with teaching us how little we know and how difficult what we do know was to come by, makes science a profoundly humbling experience.

As Max Delbrück, the physicist turned biologist who is generally considered the inspired founder of molecular biology, said on the occasion of receiving the Nobel Prize,

> The books of the great scientists are gathering dust on the shelves of learned libraries. And rightly so. The scientist addresses an infinitesimal audience of fellow composers. His message is not devoid of universality

> but its universality is disembodied and anonymous. While the artist's communication is linked forever with its original form, that of the scientist is modified, amplified, fused with the ideas and results of others, and melts into the stream of knowledge and ideas which forms our culture. The scientist has in common with the artist only this: that he can find no better retreat from the world than his work and also no stronger link with the world than his work.[6]

Once on a summer evening, I stood with a friend watching fireflies light up a field. Moved by the beauty of this commonplace event, I said, "You know, we now know what happens in the firefly's tail to make that light." Expecting a sign of curiosity, I got instead, "I enjoy beauty as it is—don't spoil it by explaining it!"

As children, we all possess a natural, uninhibited curiosity, a hunger for explanation, which seems to die slowly as we age—suppressed, I suppose, by the high value we place on conformity and by the need not to appear ignorant. It betokens a conviction that somehow science is innately incomprehensible. It precludes reaching deeper, thereby denying the profound truth that understanding enriches experience, that explanation vastly enhances the beauty of the natural world in the eye of the beholder.

When I discovered the initial step in the synthesis of proteins, I was bowled over by the beauty of the process by which all living things use energy for the construction of their substance. Soon after that discovery, William McElroy of Johns Hopkins University, in Baltimore, who was investigating the nature of the reactions that produce the firefly's light, found that the initial step in that reaction was identical to the one I found for protein construction. The same first step for building living substance and for bioillumination—an astonishing connection!

Again, when Paul Zamecnik and I discovered transfer RNA, it emerged as the *physical connection* between DNA, the source of all genetic information, and the machinery that translated that information into a new living being. That Francis Crick had imagined the existence of the connection in advance of our discovery was sheer poetry.

It is often the scientist's experience that he senses the nearness of truth when such connections are envisioned. A connection is a step toward simplification, unification. Simplicity is indeed often the sign of truth and a criterion of beauty. Scientific truth may be complex: the protein synthesis machinery of cells involves the interplay of hundreds of separate proteins and other complex molecules. But when these molecules become meaningfully interrelated in a mechanism serving all of life, they take on the simple beauty of explanatory law. As Jacob Bronowski said, "It is the sweeping simplicity of [nature's] means that overwhelms [the scientist] with a sense of awe. This is what makes nature beautiful . . . the simplicity of the materials which make so many patterns, the unity under the surface chaos. Unity is the scientist's definition of beauty, and it makes nature beautiful to him all his life."[7]

The aesthetics of explanation in biology—the view that knowledge of underlying mechanisms of living processes enhances their beauty—is most dramatically and grandly brought home by the study of evolution, the theme that has woven its way into the fabric of all biology. Evolution unifies the study of life by showing continuity, connection, and theme among the array of forms and functions in the living world.

When I started out in research, it was still not thought properly scientific, at least among biochemists, to ask *why*—

why living creatures are the way they are, why they do the things they do. Such questions smacked too much of metaphysics, wandered too far beyond the bounds of objectivity. One was wise to stick to *what* and *how*. This strictly analytical approach to biochemical problems, the dissection of living systems to reveal fundamentals of structure and mechanism, has been, and continues to be, prodigiously successful.

It has revealed, for example, that most of the basic elements of structure and function of all organisms, from bacteria to humans, are remarkably similar—in some cases, identical. We all use the same sorts of proteins made up of an identical set of twenty amino acids; we all use the same nucleic acids made up of the same four bases as genetic material. We all have similar machinery for oxidizing our food and producing our energy and for doing our cellular work, including the building of ourselves. We store, replicate, and use genetic information in the same way. The genetic code, the cipher for translating inherited information into living substance, is the same in all of us. These truths are pillars of support for evolution's first premise—that we all had a common origin.

Darwin's imaginative leap linking a common origin, hereditary variation, environmental selection, and relative reproductive success was the grand justification for asking a whole new set of questions. It was a framework upon which we hung our understanding of how the great variety of living forms has arisen since the earliest bacteria appeared some three and a half billion years ago. The science of genetics came of age seventy years after Darwin to begin the process of providing an understanding of the mechanisms underlying hereditary variation.

Before the 1940s, bacteria, by far the oldest and most abundant forms of life on our planet, were thought to have unique capabilities that accounted for their variation and environmental adaptability. This was largely because they seemed very different from higher forms: they did everything much faster, and we knew very little about their genetics. Then, in the 1940s, it was learned that the inheritance of bacterial traits could be altered by exposing living bacteria to materials extracted from dead bacteria, material that later was identified as the chemical DNA. Genes, once considered nebulous "units" of inheritance, were soon discovered to be stretches of prodigiously long DNA chains, each gene responsible for making one protein molecule. Furthermore, it was unequivocally established that the variation and adaptation of bacteria resulted from the environmental selection of rare mutant forms, as in all other forms of life. Later, when bacteria were found to conjugate, to mate sexually and thereby recombine genetic material by mechanisms fundamentally similar in all living forms, they were seen to be part of the living continuum. Their simplicity and rapid growth made them ideal experimental models. One could observe evolution in the test tube on a shortened time scale. Since bacteria were haploid, meaning that during the major part of their reproductive life cycle their genes existed in single copies, rather than in duplicate as in most other life forms, there were no problems related to hidden, or recessive, traits. Their genetics could therefore be much more easily and precisely analyzed.

With bacteria as models, and with growing sophistication in the study of cultured cells from higher organisms, biochemistry and molecular biology proceeded to reveal the physical and chemical bases of hereditary and variation in minute detail. The revelation of the structure and mode of replication of DNA, and the mechanism by which its information is translated into protein molecules, made Lamarck's theory that acquired characteristics could be passed along to offspring highly improbable.

The more we learned, the more apparent it became that

variation, obvious at the organismic level, was even more extensive at the molecular level. Variation, seen as change in an organism's protein composition upon which forces in its environment can act, was shown to be due to specific chemical alterations of information in DNA molecules that resulted in precisely predictable alterations in protein structure. Such changes in DNA were found to be of great variety, including mechanisms for substantially expanding the number of an organism's genes and for passing genes from one organism to another, and they could thus account not only for change but also for a gradual increase in complexity of organisms.

So, a century after Darwin had laid out his theory of natural selection, the molecular mechanisms needed to account for the gradual evolutionary changes in organisms and for the appearance of new species were revealed. Even more directly, molecular genetics showed that as organisms diverge in form and function, they similarly diverge in the sequences of their DNA and protein molecules. Indeed, these chemical criteria are much more accurate measures of divergence and relatedness than are anatomical criteria. For example, recent evidence that the DNA of humans and of chimpanzees is 99 percent identical suggests that many physical differences between organisms may be due to changes in genes regulating the expression of other genes, rather than to differences in the genes themselves.

The story of man's rapidly expanding understanding of his origins, beginning with inspired leaps from anatomy and paleontology through genetics to the physicochemical foundations of nucleic acid and protein structure, is one of the greatest sagas of human creativity—of the use of imagination in the rigorous pursuit of evidence to bring connection, continuity, and community to a bewildering mass of biological data. Explanation has brought harmony, pattern, meaning, and beauty in its wake. It speaks for the first time in all of man's metaphysical wanderings, directly and without maudlin religiosity or intellectual hypocrisy, to what we really are, where we came from, and how we relate to our fellow creatures.

TOWARD THE HABIT OF TRUTH

1

✣ ✣ ✣

A New Research Institute

On a rocky bluff overlooking the sea near my family's summer cottage on Cape Ann stood a large, forbidding gray house. My younger brother and sisters and I, preteen-agers, would watch the comings and goings of the occupants with uneasy awe, particularly when The President was in view. That portly, somber man was Wallace W. Atwood, president of Clark University, in Worcester, Massachusetts. Our father was a professor of physiology there, and the two men were not on good terms. Stories of the courageous struggle of the one against the administrative depredations of the other filled our home. Atwood was for us children a menacing presence.

Wallace Atwood was known as a competent physical geographer, having come from Harvard, where he had been a professor in that field. He was the senior author of a widely used textbook on geography but apparently had never been much taken with science. After coming to Clark in 1920, he built up the geography department with a heavy hand, alienating and causing to resign many of Clark's faculty in psychology, history, chemistry, and biology—areas in which the small university had enjoyed considerable distinction. He hired my father in 1931, at the remarkably early age of thirty, to be professor of physiology and head of Clark's decimated Department of Biology. Atwood and my father seemed never to have hit it off. Not only was my father young for his job; he was also ambitious, impatient, bright, and short on tact. He had, moreover, an incisive wit, great personal charm, and a leaning toward histrionics. I suspect that whatever residual appreciation Atwood might have had for science evaporated as he watched my father build his department. (That the Atwoods and the Hoaglands had neighboring summer homes on Cape Ann was, as far as I know, coincidental.)

In later years, my father was to look back with equanimity on life with Atwood, thanks to a natural mellowing and a realization that the unpleasantness and constraints at Clark eventually drove him to accomplish his greatest work: the establishment of a new research institute.

This story begins with my father because we shared a love of science and because my interest in it began with him, if somewhat perversely. While I tried hard during my schooling and early professional life to dissociate myself from him—for reasons I have found it fruitless to try to fathom—I have gradually recognized that he greatly influenced my choice of career and many of my attitudes, attributes, and antics. We were unusually close in age, he being only twenty-one years older than I. This meant that in some ways, especially professionally, we shared a generation, like brothers. But the closeness of our ages and our different personalities placed us in competition and made us warily circle one another throughout our lives. In the second half of my career, I came to share my identity with the institution he founded, and this brought a kind of intimacy that both of us cherished.

The genesis of the Worcester Foundation for Experimental Biology was unique in the history of American science. The institution sprang up as an independent basic biomedical research center without hospital or university affiliation in a community where there was no precedent or tradition for supporting pure science outside an academic setting. It appeared at a time when government funds for the support of biomedical research flowed in a bare trickle. The Foundation was the brainchild of two friends who were devoted to science, impatient with the limitations of the university research environment, and eager to create a special milieu for the conduct of free-wheeling biological investigation. Over the nearly half a century since its birth, the Foundation has contributed prodigiously to our knowledge of living processes and to human welfare. It has served as a prototype for many other research institutes, standing as a model for a more focused, efficient, collegial, and generally felicitous way of nurturing basic science in America.

Gregory Pincus and Hudson Hoagland, ca. 1965 *(Marvin Richmond, Worcester)*

The story of Hudson Hoagland and Gregory Pincus's creation of the Worcester Foundation has been told in a memoir by my father and in numerous articles and books, where it is hidden among pages devoted mostly to the development of the first and still most widely used oral contraceptive, the institution's most brightly spotlighted accomplishment.[1] From my twentieth year on, the institution impinged on my consciousness as a vague background presence, a peripheral irritant, an occasional embarrassment, a nagging nidus of guilt, and, finally, a source of uncommon satisfaction and pride.

The dominant themes of the Foundation's saga were an enduring friendship between two men and a rollicking joy in convention busting and in scientific exploration. It began when both men were graduate students in general physiology in Harvard's Department of Biology. My father's primary interest was in neurophysiology; Gregory Pincus's, in genetics and animal reproduction. They both got their degrees in 1927 and then went their separate ways on postdoctoral fellowships. Pincus journeyed to Cambridge, England, and from there to Germany, returning to Harvard in 1930 as an assistant professor of physiology. My father remained at Harvard until 1930 and then spent a year in England (1930–31) working with Lord Adrian, then Royal Society professor of physiology at Cambridge University. He moved to Clark in 1931.

Clark University had had, before Atwood's days, an exceptionally distinguished department of psychology, with a leaning toward physiology. It was hoped that my father's research interests in neurophysiology would complement and extend Clark's special strength in psychology. But the biology department was small at the time of my father's arrival, manned by one overburdened faculty member and his graduate student, who together carried the full load of undergraduate teaching. Furthermore, my father's relative youth and success in wangling an appointment that allowed him to concentrate on his research and the teaching of graduate students caused resentment among other members of the faculty. This was not helped by my father's inclination to see the Clark position as a stepping-stone back to Harvard. In addition, times were hard. It was the height of the depression, and funds for *any* purpose were hard to come by.

My father showed remarkable talent early on for garnering support for his research. In fact, he had arrived at Clark with support from two private foundations. This allowed him to hire assistants and do the research he wanted to right from the start. In addition, my mother's adoptive parents were wealthy and very supportive of my father's career goals. So, he carried with him the confidence of those who combine serious career goals with freedom from personal financial concerns. All of these elements, plus the president's lack of interest in biological science, created a formula for academic indigestion.

Nevertheless, my father proceeded to develop a productive research program, with a principal focus on sensory neurophysiology. He sought to understand the electrical mechanisms in the nervous system that allow animals to respond to changes in their environment. In 1935, he published a book, reviewing his work and that of others, on what he called chemical pacemakers: cellular reactions that appeared to be critical in regulating the rate of rhythmic events in the nervous system—heart beat, respiration rate, frequency of flashing of fireflies—and other aspects of behavior.[2] He was among the first to record brain waves in human subjects by electroencephalography, and he vigorously collaborated with other pioneers in that field. He applied the technique to patients with general paresis (of tertiary syphilis) and schizophrenia at Worcester State Hospital. The electroencephalography laboratory he established there was one of the first of its kind in the country and came to be used extensively for diagnostic and research purposes.

Gregory Pincus joined the Clark faculty in 1938. I remember my father's outrage and delight on learning that his friend's appointment at Harvard was not to be renewed after two three-year terms as assistant professor: outrage because he suspected anti-Semitism, delight because he saw a chance to bring Pincus to Worcester. Clark gave Pincus an appointment as research professor of experimental zoology and some laboratory space in a basement. Funds for his work were not forthcoming from Clark, but my father got help from two philanthropists—Lord Rothschild of Cambridge, England, and Henry Ittleson of New York. The latter was to continue to be a generous contributor to both men's work for many years. At that time, an annual budget of $5,000 would cover the salaries of a faculty member and of an assistant, as well as the costs of research. Support for Pincus initially was for only two years, but the success of the two scientists in obtaining funds planted the germ of awareness that "soft" money—nonendowment income—could be obtained by eloquent begging and could sustain a lively research program.

Pincus's interests in hormones and the chemical mechanisms by which they regulate bodily functions and my father's in neurophysiology were comfortably complementary. Their interests were to merge in a joint focus on stress and combat fatigue, particularly during the war years, and later on mental disease, especially schizophrenia. After the war, Pincus and a growing cadre of his associates devoted increasing effort to studies on mammalian reproduction and on the biological synthesis, breakdown, and mechanism of the action of sex hormones and the hormones secreted by the adrenal glands.

By 1943, the group in biology numbered a dozen out of a total Clark faculty of fewer than sixty! My father was becoming steadily more proficient at finding money to sustain research efforts. Essentially all salaries except his own, and most of the cost of equipment and supplies, had to be raised. Few of the group had regular faculty appointments at Clark. My father later wrote,

> It seemed [to Clark] that the tail was wagging the dog. . . . As time went on friction became acute between President Atwood and me. It had apparently not been his intention to have biology grow and become as strong as it was. I am sure, however, that my sometimes arrogant drive to build up my department must have been most exasperating to him and we came to hard words from time to time. Since I was a professor with tenure, I was not concerned about losing my job, although I felt unhappy as things developed from 1938 on.[3]

As a teenager growing up during my father's Clark years, I observed enough to be able to assert that this was an understatement. My father *raged* at the academic system in general and at Atwood in particular. The image of a professional environment where people with shared interests worked together in a spirit of camaraderie, free of administrative and fiscal hassles, hovered ever more enticingly in the minds of Pincus and my father.

In 1943 my father received a Guggenheim Fellowship, which relieved the university of the need to pay his salary and which freed him of his routine duties, so he could devote himself full-time to research. Pincus's salary was already covered by grants that the two had obtained. The other young associates, graduate students, and research assistants in the group were operating on a budget of not much more than $100,000 a year. (Today, fives times that amount would be needed to support such a group.) Because of space limitations, furthermore, Pincus's program had been moved into a large barn adjacent to the house my family rented on the Clark campus. My father had raised from friends in the community the money necessary to convert this space to laboratories. Not only was the barn booming with science; soon there were scientific comings and goings below us in the basement of our own house. When home on visits from boarding school and college, I occasionally descended to watch in awe the peculiar rites of science in the making. (A decade earlier, during Prohibition, my father had used the basement for another purpose: the brewing of beer. I found that activity more to my liking, for I was then more unselfconsciously curious and was sometimes allowed to operate the bottle-capping device.)

The inevitable break came in 1944 when my father and Pincus's ever more vivid and articulated daydreams gelled as a firm determination to set up their own research institute. They were already operating almost completely independently of Clark. They knew they would have to continue grubbing for money, but they liked the idea of doing it for their own cause, free from control by a university or any other mixed-purpose entity.

The success of the venture they were about to embark upon depended heavily on the confidence each had in the other as colleague and friend. In those days, it often seemed to me my father was at his best when in the company of Gregory Pincus. They were the two musketeers: full of humor, confidence, pride, and brazenness—ready to take on anyone and anything. To those men and women who came to share their ambition as colleagues, trustees, and supporters, the two men radiated an irresistible enthusiasm and commitment. They turned many of the enlightened members of Worcester society into a cadre of biology boosters. When I came to the institution twenty-five years later, the community's wellsprings of goodwill were still fresh and deep.

The Worcester Foundation for Experimental Biology was incorporated in 1944. Worcester—an industrial community with a population of about 180,000 and home to a widely acclaimed art museum, a hundred-year-old music festival, four colleges (the University of Massachusetts Medical School and its major teaching hospital did not emerge until a quarter century later), and a tradition of philanthropic generosity—gave its name to the institution, though the latter actually came to be located in a neighboring town. The word *Foundation* had a solid, basic ring, but it later caused problems when some people reasonably supposed that the institution dispensed rather than absorbed money.

My father and Pincus became the Foundation's codirectors. They assembled an impressive board of trustees, chaired by the distinguished astronomer Harlow Shapley, an old friend of my father's from his Harvard days. Vice-chairman of the board was Rabbi Levi Olan, a revered Worcester citizen and scholar. The board also included three Nobel laureates and a highly respected group of the region's top business leaders, many of whom continued to serve for twenty years and more.

With a sense of high purpose, relief, and mutually fortified confidence, my father and Pincus resigned from Clark and threw themselves into the task of recruiting scientists and aggressively pursuing funds—from friends and neighbors, foundations and government. An estate was purchased in Shrewsbury and renovated with money raised by the trustees. The group moved there in the summer of 1945.

A visitor to the Foundation in the following year recalled later that the place was small and overcrowded:

> Its one building—a former residence—had seemed adequate for the fewer than a dozen research workers who had moved into it in 1945. But, by the time of my first visit, every room and closet, from cellars to attics, had been converted into laboratories to accom-

> modate [a rapidly expanding] staff. Bathrooms were filled with tier after tier of cages housing white mice and lettuce-munching rabbits. Even an unheated sun porch has been pressed into use as a library where researchers had to don overcoats in order to consult their reference works.[4]

My father, looking back over thirty years of the Foundation experience, wrote, "We were convinced that freedom to devote full time to research, particularly in an institution in which academic rank and competitive bickering would be reduced to a minimum, would pay off."[5] It did. He had the enormous satisfaction of knowing that he had in large measure achieved what he had set out to do. The two men were greatly helped by the parallel burgeoning national commitment to the support of science, notably in the form of the expanding National Institutes of Health (NIH), whose budget steadily increased over the period of the Foundation's most rapid growth, from the late 1940s to the mid-1960s.

As the Hoagland-Pincus adventure began to unfold, I was comfortably practicing science under the protective wing of a great university. I watched the drama as a member of the audience who personally knew some of the actors. My attention span was limited, though, and, I am ashamed to admit, I was even mildly disdainful. Absorbed in the excitement of my own research, cushioned from any financial inconveniences, and a bit of an intellectual snob, I wondered what all the fuss was about: a tiny, upstart institution would hardly add much to the research output of the nation's universities. It appeared to me on occasional visits home that the enterprise did not have many of the attributes of a scientific utopia. Indeed, the tales of faculty bickering, throat cutting, personality disorders, and various forms of chicanery—tales told by my father and Pincus in tones ranging from exasperation to hilarity—convinced me I was lucky not to be involved. I learned early that what scientists see as an ideal setting can be a can of worms to the administrator, and vice versa. For the two men and their companions in the enterprise—without university, medical school, or hospital ties, without access to the special loyalties of wealthy alumni or grateful patients, without the confidence that tradition and endowment provide—the road was rough at times. They lived regularly with the fears not only of insufficient funds for research but of the institution's foundering—fears that major universities do not know. Both my father and Pincus had already acquired some distinction in science, but many of their academic peers who served on the committees that made grant awards had doubts about the stability of their enterprise and the competence of some of the scientists they gathered about them. Where genuine concern began and where snobbery and chauvinism took over was always hard to say in those early days.

But the enterprise flourished. In the years between 1945 and 1967, when Pincus died, the staff grew from a dozen scientists and support personnel to three hundred and fifty. Of these, thirty-six were independent researchers who obtained much of their own research support from grants awarded directly to them by NIH, the National Science Foundation, or private foundations. Some forty-five others were postdoctoral fellows. Funds for research grew from approximately $100,000 in 1945 to $4.5 million in 1967. One hundred acres of land adjoining the original property was acquired, and new buildings sprouted. Starting with 4,000 square feet of research space in one, inappropriate building originally designed as a private home, the campus grew to eleven buildings with a total of 113,000 square feet of prime laboratory and ancillary space.

Scientists were drawn by the personalities and achievements of the Foundation's leaders and by the opportunity to devote themselves full-time to research. The flexibility of the administrative structure, resembling that of one big family under two benevolent fathers, had great appeal. Scientists came and stayed as long as money was available. In those halcyon days of the 1950s and early 1960s, grants were relatively easy to obtain.

But grants were not enough to keep an institution alive, to say nothing of expanding, for their allowances for indirect costs of operation fell far short of actual necessary expenditures. Money obtained from private donors, mostly in the surrounding central Massachusetts community, helped to pay these additional costs. In the early days, the scientists and their assistants doubled as janitors, grounds keepers, typists, and glassware washers. Pincus took care of the animals. My mother served as bookkeeper. My father, using a power mower donated by one of the trustees, mowed the extensive lawns around the first building. He delighted in telling of a new trustee's horror on finding the Foundation's leader tramping behind the mower one day. On the spot, he pledged the funds for the salary of a grounds keeper.

To meet the institution's operating needs on a long-term basis, a membership program was inaugurated whereby donors contributed annually, as they might to a local theater or symphony orchestra. This fund, which came to represent a steady 10 to 15 percent of the Foundation's annual income, was of enormous value because it was unrestricted as to use. The infusion of essential dollars was sustained by a major effort on my father's part to promote the institution and arrange programs for donors—a labor of love for him. He willy-nilly became the Foundation's point man, public promoter, propagandist, salesman. Periodically he would hire a fund-raising or public relations professional. Disagreements and heated clashes would invariably ensue, and the specialist would be fired. It got to be a joke in the family: whom had my father hired this month to upset him? I, too, would learn how differently the world is seen by scientists and by professional money raisers.

During these years, my father had several offers of other jobs, a couple of which he seriously explored. But, for various reasons, he stayed on in Worcester. As Pincus's fame and fortunes soared following the emergence of the Pill, he was away from the institution with increasing frequency and for longer periods. Pincus's research and his enormous scientific and entrepreneurial vitality dominated the Foundation's research. My father kept the day-to-day administrative routine going, occasionally with a wistful backward look at the days of unclouded exploration. He attracted younger colleagues who carried on work of interest to him in the chemistry and physiology of the nervous system. Altogether, over the quarter century from the Foundation's founding, its scientists published some three thousand scientific papers in neurobiology, endocrine biology, and reproductive physiology.

In 1953, Min-Chueh Chang, an associate of Pincus's and one of the Worcester Foundation's first recruits, made the critical set of discoveries that led to the contraceptive pill. As that momentous story unfolded, I was so immersed in the energetics of protein synthesis thirty miles away in Boston that it had little impact on me. The big event of that year in my realm of interest was Watson and Crick's announcement of the structure of DNA. The emergence of the Pill has since come to intrigue me, not just because it was "in the family" but also because it involved a fascinating interplay of science, social urgency, coincidence, and entrepreneurial drive.

It began with Pincus's awareness of the work of A. W.

Makepeace and colleagues published in 1937, showing that the injection of progesterone, the pregnancy hormone of the ovary, inhibited ovulation in rabbits. Chang confirmed that work in 1951 and then proceeded to test other compounds with similar properties that might be effective orally

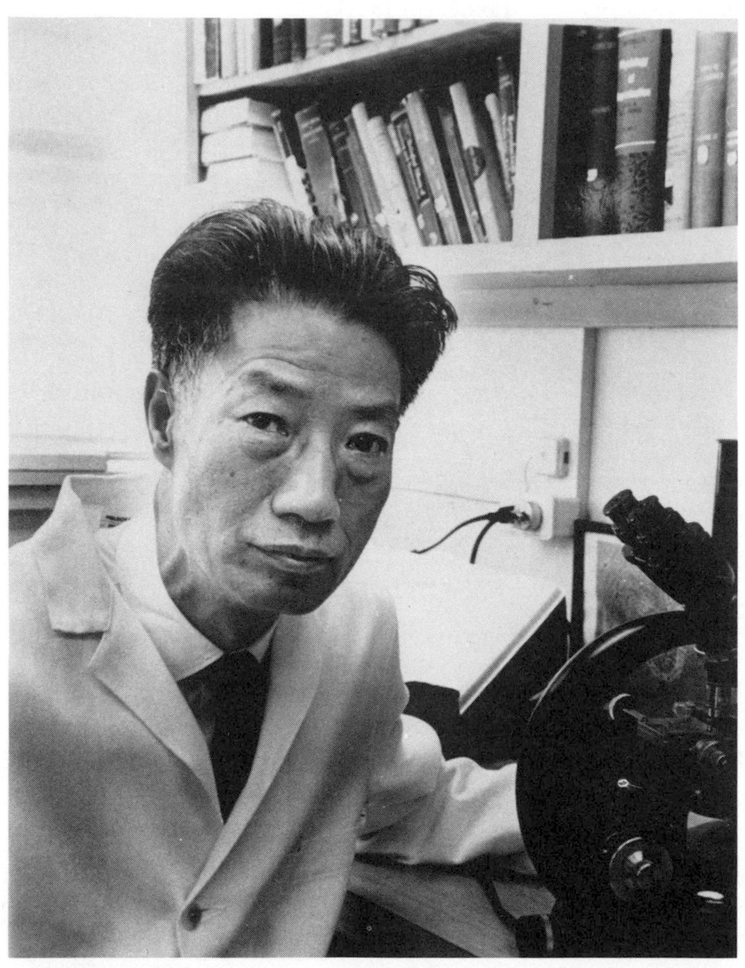

Min-Chueh Chang, ca. 1975 *(Marvin Richmond, Worcester)*

and have fewer side effects. In the meantime, John Rock at Boston's Free Hospital for Women (now a part of Brigham and Women's Hospital) had been trying to *induce* ovulation and pregnancy in sterile women by using progesterone and estrogen. He reasoned that infertility might in some cases be the result of an underdevelopment of the uterus, remediable by the administration of hormones. Pincus and Rock happened to meet at a scientific conference and to discuss their mutual, if opposite, objectives. Pincus suggested that Rock administer the progesterone for twenty days during each of a succession of menstrual cycles, then stop it and hope to produce a sort of "rebound" ovulation in the ensuing normal (untreated) cycle. This Rock did, and it appeared to work, because several of his patients were able to become pregnant. It was assumed that ovulation had been inhibited during the treatment cycles, but this was difficult to prove at the time.

Since the natural hormone progesterone is readily broken down in the gut, large quantities of it are necessary for oral administration. This made it prohibitively expensive to use as a contraceptive agent. Pincus therefore began a search for cheaper and stabler analogues—chemical variant forms that do the same job but have a slightly different structure. By virtue of one of those coincidences by which science often advances, analogues called progestins had at about that time been synthesized by pharmaceutical chemists (with no notion of their use as contraceptive agents) and were becoming available for uses similar to those envisioned by John Rock. Chang soon found that certain progestins were nearly 100 percent effective in laboratory animals as orally administerable ovulation inhibitors.

In another felicitous coincidence, my father's research at the Worcester State Hospital had brought him in contact with the Neuroendocrine Research Foundation, which had been established by Katherine McCormick to fathom schizophrenia, a disease from which her husband had suffered. Well-educated (she held a degree in biology from MIT), socially conscious, and enormously wealthy, Mrs. McCormick was interested in furthering population control and contraceptive development—concerns she shared with her friend Margaret Sanger. She learned about Pincus and Chang's work from my father, and she and Margaret Sanger met with Pincus in 1951. That meeting gave substantial impetus to Pincus's drive to focus on the contraceptive promise of his and Chang's research. It also resulted in Mrs. McCormick's generous funding of their work. Indeed, the Worcester Foundation's first and principal endowment fund was established by her.

Contraception was at that time a closet topic. It would take immense social pressure and chutzpah to bring before scientists, the medical profession, industry, and the public a chemical birth control agent. Pincus was up to the task. The International Planned Parenthood Federation, founded by Margaret Sanger, was to hold a major meeting in Tokyo in October 1955. That meeting was Pincus's critical challenge. In opposition to the views of many of his colleagues, including John Rock, who felt it was premature, Pincus and Chang determined to go to Tokyo and not only present their scientific results but also state the clear implications of their work for human contraception. Their paper got worldwide press attention and set in motion the astonishing chain of events that moved chemical contraception from extensive clinical trials, beginning in April 1956, to the acceptance of the Pill by the Food and Drug Administration in June of 1960. Pincus, John Rock, and many colleagues joined forces for the clinical trials and proceeded to make history.

(John Rock is a personal hero of mine. He and I had a special relationship: as a young physician, he assisted my mother's obstetrician in bringing me into the world. Later we became friends, and I greatly enjoyed his company.

Boisterous, cheerful, dedicated to his patients and his profession, he believed deeply in the right of women to control their own fertility. As a devout Catholic, he worked long and hard to persuade his church to sanction contraception. When, in 1968, in an uncompromising encyclical, the pope set the present church policy forbidding use of the Pill, Rock was wounded and outraged. He had been virtually certain the church would yield to the evidence that Catholics were using the Pill as frequently as non-Catholics. In the early days of the Pill's ascendancy, at the height of his fame and infamy, an angry Catholic wrote excoriating him, "You should be afraid to meet your maker." He wrote back, "My dear Madam, in my faith we are taught that the Lord is with us always. When my time comes, there will be no need for introductions.")

Two companies, G. D. Searle and Syntex, profited enormously from the Pill for having patented its synthesis (without knowledge of its contraceptive potential). Searle had provided modest assistance to Pincus in the earlier days at Clark, and he was a consultant to the company for many years thereafter. When the boom came (some sixty million women were using the Pill by the 1970s), the Foundation had no claim to any financial benefit. Over the years since Pincus's death, first my father and later I tried to persuade these companies to give the Worcester Foundation an endowed chair or other significant token of their profound indebtedness to the work of Pincus and Chang—to no avail.

By giving women a measure of control over their fertility and thus their personal destinies, oral contraceptives dramatically altered the fabric of society. The Pill also raised hopes that mankind had within its grasp a sound method for controlling the worldwide growth of population. While the Catholic church, human recalcitrance, and limitations on the usefulness of chemical agents have partly dimmed these hopes, an enormous forward step had been taken.

Pincus died in 1967 from a failure of his blood-forming tissues, apparently caused by chronic exposure to organic solvents in connection with his chemical work on hormones. He vigorously pursued his scientific interests and the promotion of the Pill until shortly before his death, remaining full of enthusiasm and humor and still offering encouragement to his younger colleagues. His death seemed to tear the heart out of the Foundation. Many of the Foundation's scientists had been drawn there by Pincus's charisma, and his death left them bewildered and exposed. My father's interest in managing the institution had been waning and now seemed to collapse. He asked to be relieved of his duties the following year. The two leaders had given the institution its vitality and structure. Nearly a quarter century of a grand new venture was ending in sadness and a sense of faltering purpose.

Fortunately, the first years following Pincus's death would prove instead to be ones of transition for the Foundation, leading to a new phase—in which I would be deeply involved.

2

✣ ✣ ✣

First Steps

One day, out of the blue, my daughter Judy telephoned and asked, "What made you become a scientist?" She caught me just as I was beginning to write this book and was shuffling evanescent memories and emotions in trying to answer that very question. She was starting to study chemistry, at the age of forty, and as a requirement for a course she was expected to interview a scientist about motivations and creativity in science. Judy is a bright and vibrant mother and part-time teacher. Her last contact with science had been in college, where, while majoring in liberal arts, she had tackled an advanced course in chemistry and flunked it. That failure rankled. Now she thought that chemistry could be taught better, and with an almost missionary zeal she was intent on designing a course for secondary-school students.

During our interview, she called my attention to a study of sixty-four eminent scientists that tried to evaluate factors in their backgrounds that had influenced their choice of career.[1] I paraphrase the author's picture of the "average" successful scientist:

(1) He (there were no women in the sample) was the firstborn son of a middle-class family, in which the father

was a professional man. (2) He had a high IQ and did a lot of reading at an early age. (3) In school, he was lonely, shy, and aloof from classmates. (4) He had only a moderate interest in girls and did not begin dating until college. (5) He married at twenty-seven and had two children and a relatively stable marriage. (6) He decided to become a scientist as a result of an opportunity to do a piece of independent research and, on discovering the pleasure of research, never turned back. (7) He avoided social affairs and political activity. And (8) religion played no part in his life.

Judy and I agreed that only 1, 6, and 8 fit my experience, giving me a sorry score of 38 percent!

I had to say that the immediate and obvious cause of my commitment to science was having the good fortune to stumble upon the right sort of problem in the right milieu at the right time. But what had prepared me to respond to that challenge when it came? It was apparent to both of us that I had been influenced by my father. I *was* the firstborn son of a scientist, although I was never conscious of having had a real interest in his work before I got into science myself.

Growing up, my two younger sisters, my brother, and I were much on our own or in the care of a housekeeper in a big, busy, socially lively household. Both our parents often acted as though we children were unexpected, but not necessarily unwanted, guests. My mother was beautiful, amusing, wistfully sad, and musical. Once asked by a teacher why his mother didn't make him wear warm boots to school, my brother replied, "She was playing the harp." My father obviously loved his work. He left for his laboratory early in the morning and appeared for supper on the run, returning to his work to be there long after we children were asleep. His passion for his work was, if anything, a factor in convincing me early on that I did *not* want to be a scientist. Yet the daily evidence of his dedication, enthusiasm, and especially the freedom to do the things he wanted to do was persuasive, even to an obtuse teenager. Indeed, it never occurred to me that I would *not* enter a profession. He open-mindedly insisted that I should do anything I wanted when I grew up, but always with the proviso that I do it better than anyone else—a heavy burden for a floundering teenager. And he did convey to us children his deep conviction that we had an obligation to give something of ourselves to society.

My father was constantly in motion. He had an easily ignited and quenched temper and a showman's antic sense of humor. I remember a photograph of him as a young instructor at

Anna Plummer Hoagland, ca. 1916

Cambridge University wearing a medieval knight's helmet (he collected ancient armor and weapons) and preparing, in a Saint George–like stance, to plunge an enormous two-handed sword into a fallen bicycle. Occasionally he terrified visitors at our house by greeting them from the top of the stairs, feigning dizziness, and tumbling head over heels down the stairs to lie inert at their feet.

He was forever trying new things. Once he brought to our summer home some immense balsa wood gliders, which we children launched in the dunes and chased for what seemed miles—joyously amazed that they could defy gravity so long. On another occasion, he arrived with an Eskimo kayak on the top of his car, carried it down to the beach, launched it, firmly tied the drawstrings of the canvas cowl around his waist, and shouted to my mother and us children to watch him do an Eskimo roll. One hundred feet from shore, he gave an immense heave with the paddle and rolled 180 degrees. He didn't reappear. Seconds seemed like minutes. A friend of my mother's became alarmed and said, "We should help him; he's stuck." My mother: "Don't be silly, he's just showing off." Friend: "I don't care, I'm going out there." She saved his life. Hanging upside down under the kayak, he'd been unsuccessfully struggling to untie the knot that held him in.

My father's attempts to interest me in his work were sporadic and awkward, and I tended to respond with teenage indifference, but there were magic moments when all the pieces fit into place. I remember a long ago walk with him on a lonely beach. The sea was gray; ragged clouds scudded before a chill, early winter wind. It was a day for discovery. Lying in the decaying seaweed at the high-tide border were washed-up, empty bottles of all shapes and sizes. It gradually dawned on us that every bottle was capped: we couldn't find a single topless bottle. We were puzzled by such unwonted conformity among bottles until my father struck upon the explanation. He urged me to see in the bottles a grander meaning. The upshot was a lesson in evolution, firmly screwed into my head for life, because those bottles were obviously the few survivors of an ocean journey, the fit few. Of the many empty bottles thrown into the sea by the hand of man, a tiny fraction would have had their tops replaced by some inadvertent, chance act, rendering them unsinkable. The vastly larger number of nonsurvivors caplessly ill-suited to the ocean's hostility, would have long since sunk to the bottom.

Two other linked events in my early life with my father have a kind of symbolism. When I was seven, my father, still in the biology department at Harvard, learned that the university's Peabody Museum would dispose of a large, standing, stuffed tiger—a majestic beast, if somewhat the worse for wear. My father responded without delay: with the aid of museum maintenance men he loaded the animal aboard his four-door convertible Ford—front feet on the front seat, rear feet in back—and drove off through Cambridge to our house. Many people have since spoken to me of their astonishment at seeing my father proudly driving a tiger through Harvard Square. We children were ecstatic, and the tiger settled in as a fixture in our home for years.

Not long after the tiger episode, I stole a stuffed eagle from the school I was then attending and smuggled it into my room at home. The feat was the more notable in that the bird's wings were fully outspread and I had to carry it half a mile along a busy city street. The prize stood magnificent on my bureau overnight; then my mother found it and ordered me to return it to the school.

The tiger and the eagle have become a family metaphor for my career. The eagle is the soaring of the intellect that is science. Throughout my career I tended to view myself as an amateur among professionals—an interloper, one who came to possess science, as I did the eagle, joyously but

Mahlon Hoagland, ca. 1927

illicitly and all too briefly. And I strove, if subconsciously, to emulate my scientist father.

Although my father and I dueled throughout most of our professional and private lives, the weapons were blunted by a sort of wary affection and a lot of kidding and humor. He was always interesting and provocative and responsive. An enthusiastic humanist committed to liberal causes, he worked with dedication with laymen, philosophers, and religious leaders to plead for rationality in the conduct of human affairs and in the furtherance of world peace. He was particularly fond of the American Academy of Arts and Sciences, of which he was president from 1961 to 1964. This association afforded him an unusual opportunity for communication among scholars who had wide-ranging societal interests. When he died, at the age of eighty-two, after our lives had converged and we had had twelve years together as fellow professionals, I had occasion to say of him with deep conviction, "He did in big measure what most of us, in one way or another, want to do. He put much more of himself into the flowing stream of the generations than he took out."

My career in science has been marked by two special circumstances. First, it coincided with and permitted me to participate in the rise of molecular biology—that burst of discovery between the mid-1940s and the early 1960s that revealed the nature of the gene and its mode of expression. Second, unlike the vast majority of my colleagues who practiced science in universities, I spent twenty-seven of my thirty-six years in science in research institute settings. My father had a role in creating both circumstances. He conspired with my mother to conceive me at the right time;

his professional example drew me, albeit kicking and protesting, toward science; and his impatience with academia and his propensity to seek an ideal environment for research were in some mysterious way conveyed to me.

Late in high school, I decided I wanted to go into the practice of medicine. My reasons were rather vague: there was the romantic appeal of a career of service and a desire to avoid direct competition with my father. When, after medical school, I almost accidentally confronted a scientific problem that required my full attention, I discovered, as if by revelation, that science offered me an unprecedented degree of freedom to define and pursue a career *on my own terms*.

In my primary-school years, the prospect of my becoming a professional person seemed dim. When I was thirteen or fourteen, the principal of the school I attended told my parents that neither I nor my younger brother was college material. (My brother attended college and was a successful businessman at the time of his accidental death at age thirty.) I spent two years in public high school never getting out of academic low gear and was then sent away to a private school, where I did my first scholarly work largely in emulation of the friends I chose.

World War II broke out as I began college. I had chosen Williams College but transferred to Harvard after a year. Harvard was a somber place in 1941, when I entered; I became immersed in an accelerated, no-vacation, no-nonsense curriculum punctuated by the departure of friends volunteering to go to war. My undergraduate time there was brightened by my meeting and marrying my first wife, Elizabeth Stratton.

In my home, pacifism had been articulated regularly and forcefully. I had been shown picture books of the horrors of trench warfare in World War I and admonished "Never Again." But the Battle of Britain and the overriding need to stop Hitler changed all that as Roosevelt, whom most of us ardently supported, moved the nation toward helping our allies. My patriotism never got sufficiently honed, however, to persuade me to volunteer for military service. I defiantly, if somewhat guiltily, vowed I was not going to leap into the breach before I was drafted. And I planned to go on to medical school, after which I would be obligated to serve in the armed forces anyway.

I pursued premedical studies, working below what I and others knew to be my capacity. I was troubled by my lack of ambition. Occasionally, to my surprise, I was charmed by good teaching into exploring a subject in depth. Professor Louis Fieser's course in organic chemistry had that effect. I was enthralled with his ability to breathe life into molecules and make them big enough and palpable enough to require his wrestling with them on the lecture hall bench. For a period of several weeks my roommate and I spent our evenings constructing implausible organic syntheses that required two-meter lengths of glued-together sheets of paper—just for the fun of it.

The wartime urgency to train doctors pushed me, without a degree, across the Charles River to Harvard Medical School in June of 1943, after a year and a half at the undergraduate college. The war had compressed the total time between the start of college and the completion of medical school from the usual eight years to five and a half. Admission was easier than I had expected: Harvard was then less attentive to quantifiable academic and personal qualities in its candidates, and the process rather resembled admission to an elite men's club. But once admitted, we plunged into an unremitting grind, cramming the work of four years into three. The curriculum seemed an accretion of poorly integrated, ponderous pedagogical burdens imposed on us by an endless procession of research-oriented faculty members, almost all of whom, regardless of ability, wanted to take a whack at teaching. This state of affairs was not—and is not—limited to wartime or to Harvard. There was no time for scholarly reflection. Even without acceleration, the intellectual integration of material into a coherent whole is difficult; in our case, it was well nigh impossible.

All medical students at that time were required to be in military service. Our classes were about evenly divided between Navy midshipmen and Army privates. We were to receive our commissions and military assignments upon graduation. Those of us who opted for the Navy were happily left alone to pursue our studies. Our Army brethren were regularly harassed by a fanatically militaristic Army sergeant who possessed not a trace of academic experience, grace, or intelligence. He required that these poor students wear uniforms and march about the campus.

In spite of the wartime grimness, medical school did open vistas for me and helped me gain self-confidence in the intellectual realm. I recognized, somewhat reluctantly at times, the legitimacy of such a high-grade trade school's demands. And most of us were committed to what we saw, without cynicism, as a noble service to mankind. Later I came to believe that my medical education provided a far better background for a life in science than did the education most of my colleagues with Ph.D.'s had received, primarily because of its greater breadth.

The science to which I was exposed during the first two years of medical school was required because of its relevance to a practical end, but I found it singularly uninspiring. I remember wanting very much to share the enthusiasm for research that some of my professors evinced, but I never succeeded. Science as I was exposed to it then never remotely resembled what I later found it to be in practice. I am sure my belated encounter with "real" science was all the more exhilarating because the nature of the experience had been so well disguised earlier.

Anatomy is one of the first courses one takes in medical school, and it generally was not fashionable among medical students to like it. I did. In fact, the content of no other single course has stayed with me so tenaciously. Perhaps that was because it afforded such a vivid encounter with the exquisite artistry of the physical self—in the guise of a subject classically, romantically, and unambiguously related to medical practice. Surgery, too, was viewed by the colleagues I liked most as intellectually inferior to internal medicine. I tried to share their prejudice, but my experience worked against it. Too often my medical professors seemed stuffy and pompous and uneasy with students.

Early on, however, I discovered and admired the work of the surgeon Robert Gross, although I never met him. His book, written with the pediatrician Ladd, on the surgical repair of congenital anomalies in children, was a classic. And I was greatly influenced by two surgeon-scientist-teachers: Ira Nathanson and Oliver Cope. Their intellectual genuineness and enthusiasm for medical exploration drew me out of what I considered the oppressiveness of academic formality. These men talked companionably to us students as though our opinions were of value. Above all, they asked questions whose answers were not known. As surgeons with inquiring minds, they had a way that seemed to me glamorous and exciting. They were the heroes I decided I had been looking for during the long, gloomy academic pull through high school, college, and medical school.

In my third year of medical school, I found I had a natural aptitude in the surgical laboratories, where we performed operations on dogs. I surprised myself by my own competence. My confidence got a further boost from the third-year course in obstetrics. I was among the last crop of students at Harvard Medical School to deliver babies "on district," meaning in the home. My first call was to a tenement in Jamaica Plain, to which I went with a one-reading-of-the-textbook background and a fellow student as assistant. We were greeted by a weary mother experiencing only an occasional mild pain, surrounded by eleven noisy offspring eagerly anticipating the upcoming event. After locking the kids out of the bedroom, we settled down for a long wait. I figured we would have plenty of time to boil

Ira Nathanson, ca. 1950

Oliver Cope, ca. 1978 *(courtesy Oliver Cope)*

the water and sterilize our rubber gloves. Abruptly our mother lay down on the bed and quietly proclaimed, "It's coming." He did, and I barely caught him as the umbilical cord broke. I clamped the cord and looked for my assistant. He was standing against the wall of the bedroom, eyes as big as saucers, pale and paralyzed. I felt great!

On my subsequent daily visits, the mother repeatedly expressed her admiration for the competence of the "Harvard doctors." I could not help being amazed at how easy it was to please a patient if you had *anything* positive to offer. And I took great satisfaction in a practical demonstration that I could do something useful on the basis of knowledge from a book. (A friend who had been in a similar emergency told me that in his panic he had reached into a pot of boiling water to don his rubber gloves. He had screamed in pain and his patient sought to give him first aid.)

I should not have been surprised by my attraction to surgery, for it had roots much earlier in my life. I used to do a lot of drawing, but the requirements to reduce three-dimensional objects to two often frustrated me. I discovered at a young age, however, that I could comfortably represent things in three dimensions in clay or wood. Sculpture—especially sculpture involving the irreversible cutting away of material to reveal something envisioned within, as opposed to molding—has been an integral part of my life ever since. To find my innate manual dexterity prized in the academic-professional setting in which I had worked, unmotivated, for so many years was enormously stimulating.

So, I set my sights on a career in surgery. The decision seemed to stem from my awareness of a natural aptitude; the autonomy implicit in a technological-artistic occupation; the attraction of work that produced unambiguous results, giving one the sense of being in control; the satisfaction of being of service; the drama and glamour; and my lack of any real insight into what the life of a surgeon would be like. It has fascinated me that all of these considerations, except the last, also applied to the career I finally chose.

I had looked forward to the clinical experience in medical school, but when it came I was disappointed. With an irritation that probably hid a lack of self-confidence at the bedside of miserable people, I deplored my fellow students' apparent preoccupation with the disease instead of the patient. And much of the daily routine of the care of the sick had little appeal. Dr. Rita Kelley, an admired colleague of later years and an outstanding Boston internist, remembered me on the wards and told me years later she had bet I would never be a clinician. She said I seemed bored. In spite of my lack of an absorbing interest in disease as such, however, I still was not deterred from seeing surgery as a service, whose practice would be enormously satisfying.

It has seemed to me, looking back, that I have been inordinately lucky most of my life. I have been vigorously healthy and had enough money, a variety of absorbing interests, and good companions, and the timing of my confrontation with significant scientific problems has been propitious. But while in medical school, within three months of getting my M.D. degree, I received a frightful blow. In October 1945, I was discovered to have tuberculosis, apparently picked up from an infant with widespread (miliary) tuberculosis whom I had cared for on the ward at Children's Hospital before the severity of her illness was appreciated. Of all the diseases we had learned about, tuberculosis was the one we feared most. Doctors, nurses, and medical students were sitting ducks for the disease, especially during the war years when many of us were overtired, overworked, and worried.

The discovery of my illness was followed by three nightmarish months, during which I progressed from "minimal" to "moderately advanced" tuberculosis on a ward at Chelsea Naval Hospital, where I was sent because I was a midshipman in the Navy. It was then near the end of the war, and large numbers of men were returning from the European and Pacific theaters with severe pulmonary infections, often undiagnosed. I was on a fifty-bed ward with many extremely ill men, several of whom died during my stay. Separated from my family, frightened and deeply depressed, unable to see any way out of the military morass, I was sure I, too, was going to die on that ward. My life was literally saved by the sensitivity and pertinacity of Dr. Walter Burrage—a prominent and widely respected Boston physician and a Navy captain at the time, about to be mustered out of the service—who discovered me and slashed a good deal of red tape to get me into private care.

I was sent to Trudeau Sanatorium, in Saranac Lake, New York, for the classic "cure." This was before the availability of chemotherapy, and I was put on bed rest twenty-four

Walter Burrage, ca. 1970 *(courtesy Katherine Burrage)*

hours a day. The first couple of months were bleak and terrifying. With a cavity in one lung and an additional diagnosis of intestinal tuberculosis, for which the prognosis was grim, I was pretty sure my tale was told. The latter diagnosis proved to be in error, however, and over the next nine months I steadily moved from horizontal to vertical, my spirits following suit.

The experience of sanatorium life was strange and wonderful, as Thomas Mann told in *The Magic Mountain*, a book I devoured in my first weeks at Trudeau. (The Trudeau librarian, a tuberculosis "veteran" of fifteen years, would not let new patients read *The Magic Mountain*, deeming it too depressing and therefore bad for them. I had to go to the institution's director to get permission to read it.) I, too, discovered a paradox that Mann had described: when one's life's progress is measured by monthly chest X rays, time passes very rapidly. During those speeding hours, I read widely and got to know my fellow patients, thus experiencing the liberal education I had missed in my rush through college and medical school.

As I look back on that struggle with the tubercle bacillus, however, I can see that its most lasting impact was its putting me in the front lines when chemotherapy for the disease was made available in early 1946. Many of my friends at Trudeau were among the first to be treated with streptomycin. I was responding well to bed rest and did not receive the antibiotic treatment. Those who did receive streptomycin were patients who had advanced disease; some of them had been hospitalized for twenty years. In my final months at Trudeau, I witnessed the daily miracle of their return to physical health and shared the joy and anguish of their struggles to grasp the implications of beginning to live normally again. It was an indelible reminder of the impact of basic research on medical progress. I returned to Trudeau a few years ago for a meeting. It now is a research

Joseph Aub, ca. 1950

institute, the sanatorium having closed its doors in 1954. I asked to see the cottage that had been my home for nine months. Only the cement corner posts remained, hidden in weeds.

When I returned to medical school to repeat my fourth year after a two-year absence, I felt a lot older, very much alone, and still bent on a career in surgery. Indeed, I was scheduled to begin an internship at the Massachusetts General Hospital, after my delayed graduation in 1948. When the chief of surgery, Dr. Peter Churchill, who had been on leave during the war, resumed his duties and found the hospital about to take on a recent victim of tuberculosis, he summarily canceled my appointment. Angry and not a little scared, I applied at the Peter Bent Brigham Hospital and was accepted. After only a few weeks into that internship, an X ray showed that I had reactivated my tuberculosis. My body appeared to be trying to tell me something, and I reluctantly responded by looking for a job less physically demanding.

In the summer of 1948, I applied to Dr. Joseph Aub, director of the Huntington laboratories (the John Collins Warren Laboratories of the Huntington Memorial Hospital of Harvard University at the Massachusetts General Hospital), who promised me a postdoctoral position in the fall. When I reported for work, he had forgotten the agreement but assured me I could work on any number of interesting projects. To this day, I marvel at both the casualness with which I obtained my first job and my immense good fortune in getting it.

Joseph Aub was a physician-scientist of extraordinary versatility. His investigations of lead poisoning, showing that the metabolism of lead mimicked that of calcium, that the same hormone that promoted calcium excretion from the body (the parathyroid hormone) also promoted excretion of lead, and that calcium could relieve symptoms of lead poisoning were of fundamental importance. Aub became director of the Collis P. Huntington Memorial Hospital in 1928. That hospital had been established under the Harvard Cancer Commission to treat and do research on cancer. It was one of the first places in the United States to use both X rays and radium in cancer therapy and had installed the first million-volt X-ray equipment. Huntington surgeons invented the electro-cauterizing knife, and researchers there pioneered the development of high- and low-incidence cancer strains of mice and made substantial contributions toward understanding genetic factors in cancer in laboratory animals.

Aub committed the Huntington to fundamental investigations of the biology of normal and abnormal growth. He set a course that broadly attacked cancer as an aberration of growth control, a visionary and radical approach at the time. It was not until the 1950s that it became respectable in cancer research circles to see basic biological research as relevant to an understanding of cancer. Ira Nathanson, who had influenced me as a medical student, was one of Aub's associates, and together they made important observations on the effects of estrogens on the progress of breast cancer, achieving dramatic remissions in some postmenopausal women with surgical removal of the ovaries, the source of these hormones. In addition, Lewis Engel, another member of the group, for many years made the Huntington a center for the study of the metabolism of sex and adrenal gland hormones in cancer patients. Paul Zamecnik, soon my mentor and lifelong friend, began his long association with Aub and the Huntington in 1938, investigating the biochemistry of cell growth.

It was never easy to get money for operational costs, and in 1942 the Huntington Hospital closed shop. Its outpatient department merged with the MGH's tumor clinic, and the laboratories became a part of the MGH.

Paul Zamecnik later succeeded Aub as the Huntington's director and as professor of oncologic medicine at Harvard. He said of Aub at the time of his retirement,

> Of his generation in medicine he could say with Virgil, "These things I saw and part of them I was." He sat with grace at all tables; nourished ugly ducklings while their feathers were changing; served as family doctor and confidant to many colleagues; fought a gallant, losing battle against the laboratory coffee break;

> gave up smoking many times; and was often late but on entry never failed to brighten a dull room.[2]
>
> We all came close to worshiping Joseph Aub.[3]

The Huntington opened the door to science for me by way of a project close to Aub's own interest in toxicology, and I will relate more of that adventure in the next chapter. Here I will simply note that in the course of that project I found myself. It seemed to me that all the pieces of my personal puzzle—family, environment, schooling, and yearning to immerse myself in a meaningful role—fell into place. The appeal of the experience lay in two vivid and unexpected qualities: its freedom and its drama. The freedom was the more obvious: *it was all up to me.* Implicit in the freedom, too, was the exhilarating sense of having control over a powerful process by which one made order from chaos and provided explanation where none had existed. I was being offered an opportunity to embark on an expedition of exploration, to conduct it myself, and to define its objectives and limits—all expenses paid! Surprisingly, my education licensed me to undertake this fantastic venture. There was freedom in another sense: I was liberated from academia, books, courses, pontificating professors, mountains of irrelevancies. I could go back to whatever I needed of all that, but the bulk of it was joyously behind me. (François Jacob, a leading figure in contemporary biology who came somewhat late to the field, recalled wondering whether he should take some courses to update himself. His mentor André Lwoff advised him, "Do experiments, and the rest will come on its own. If you insist, go after one or two certificates, biochemistry or genetics. But on the side, extra. Don't let it keep you from working. . . . Research is mainly a matter of flair."[4])

My sense of the drama or theatricality of the experience is more difficult to explain. It derived from a peculiar, oddball view of science as theater that I can clarify only by another look backward. From an early age, I had a sense of being a performer, of being partly outside myself looking on and wondering how others saw me. My parents encouraged, even applauded, certain vaudevillian tendencies that appeared early and persisted despite the requirement to consider more serious pursuits. I sang, played the guitar, hammed up Robert Service poems, and even founded an acting company that produced melodramas. (I was the villain three years in a row. Kids used to point me out on the street and feign fear.) I envisioned a career as a kind of role-playing. I went to medical school with an uncompromising if naïve vision of becoming a hero in the service of mankind. I had responded to the standard stimuli: Hans Zinsser's *Rats, Lice and History,* Paul de Kruif's *Microbe Hunters,* and Sinclair Lewis's *Arrowsmith.* My idealism was only slightly diminished by the uninspiring curriculum of medical school and my disillusionment with the mundane details of caring for the sick. My plan to be a surgeon was in a sense a showman's choice.

Backing into science at the Huntington after my clash with tuberculosis—itself a romantic disease in fiction—had all the drama of revelation. I turned around and was beguiled by the freedom and power, removed quite completely from the everyday, mercantile world. I reveled in the clear, epic vista that research opened before me to work out my own way of alleviating some of the major afflictions of mankind.

The metaphor of science as dramatic performance goes beyond my own, sentimental view of my career. Scientists deal with *things*—measurable, controllable, predictable—much more than they do with people. Like actors and artists—musicians, painters, sculptors—they devote much of their lives to exercising their imaginations and innovative skills and to interacting with and manipulating their materials to make special things happen. They do prance upon a stage, once removed from the human struggle. They act out discovery. They regularly jolt their audiences with novelty with which *they*, the audience, must deal. And like actors, scientists remain somewhat mysterious and controversial—and aloof.

What gives the drama of science its greatest force is that the revelations of its actors are real. Its insights are irreversibly intrusive—in society, in our personal lives. Science is a process, a progression, because its discoveries are cumulative, the stones of an edifice under construction. While science gives us reliable knowledge about the external world, it cannot do much directly to solve problems of values that press upon us in our daily lives and that have no objective solution. This should not preclude the use in our daily lives of the scientific habit of skepticism, its esteem for evidence, its awareness of how hard it is to get at the truth, its insistence on distinguishing between testable hypothesis and belief.

I did not see all this at the time, but I saw enough to become enthralled.

When I was seventeen, I worked in the summer as a paid hand at a remote camp in a vast timberland in the Adirondacks. On a lone and listless walk one day, I stumbled on an overgrown logging railway and, on the track, an abandoned handcar. With awakened zeal, as though the vehicle had been left for my private use, I leapt aboard and pumped off into the forest. After covering half a mile, I found myself coasting down a hill to the side of which, in the valley ahead, was a small lake dotted with what appeared to be hummocks sprouting dead branches. But as I clattered nearer, the hummocks began to move away from me—they were all swimming deer. I rode past, entranced by the wonder of the scene. At dusk I left the handcar on the track, and I see it now, silent in shadow. That astonishing discovery of a track and a vehicle, laid out by others who came before me and enabling me to plunge into unimagined realms of beauty, took on, with time, the quality of a metaphorical glimpse of my career in science.

3

✣ ✣ ✣

Vistas

In 1947, a boy rummaging in the trash cans behind the Massachusetts General Hospital found a spent fluorescent light tube and used it as a bat to hit a ball. The tube shattered, and fragments of glass punctured the skin of his face and arms. He was released after treatment in the emergency ward but returned months later with hard lumps at the points where the phosphor-coated glass had penetrated his skin. Biopsy showed the lumps to be granulomas—hard masses of cells and connective tissue. The lesions resembled those seen in the lungs of women who had been employed in the fluorescent lamp industry in the 1930s and 1940s. These women salvaged the phosphor from used light tubes by running an electric-powered wire brush inside the tube. The air they breathed was thick with the fine dust. Years after this exposure, many of them died with widespread granulomas of the lung.

In those days, phosphors contained the metal beryllium—in the form of insoluble beryllium oxide, along with other metal oxides—and there was growing suspicion among toxicologists that beryllium was the agent responsible for lung disease. By the late 1940s, the fluorescent lamp industry, under increasing pressure from medical investigators, was reported to have stopped using beryllium. But the metal remained important in the nuclear industry, in the making of heat resistant materials, and in certain other processes. In 1946, it was discovered by toxicological testing of beryllium compounds on animals that the intravenous injection of beryllium oxide, which concentrates in bone, caused bone tumors in rabbits. These cancers closely resembled certain highly malignant bone cancers in humans. There was also evidence that inclusion of beryllium in the diet of young rats produced a condition of the bones resembling but clearly different from rickets caused by vitamin D deficiency. Beryllium was found markedly to diminish the calcifying capacity of bone cells cultured in test tubes.

The incident of the boy and the glass bat prompted Joseph Aub and a colleague, Robert Grier, to explore the mechanism of the action of beryllium in biological systems. They suspected that the metal might interfere with the action of certain enzymes and, in particular, that alkaline phosphatase, apparently critical in the normal calcification process in bone, might be inhibited by beryllium.

Grier soon discovered that the metal did indeed interfere with the normal action of the enzyme. Together he and I showed that beryllium was a potent inhibitor of alkaline phosphatase, apparently because it competed with and took the place of magnesium, an element known to be essential for the enzyme's activity.[1]

No study of the bone cancer–inducing property of beryllium had ever been published. Because of its significance, Aub encouraged me to confirm this finding, which I did during the next year and a half.[2]

It seemed to me that in order to learn more about the chemistry underlying the biological effects of beryllium, I needed something better than a rabbit as a model. I wanted a simpler system that might respond to beryllium in a clear-cut, measurable way and in a shorter time. Being new to the game of science and jealous of my independence, I also sought a system that would be *my own,* allowing me full scope to explore it before others got into the act.

I unearthed a paper showing that beryllium dramatically inhibited regeneration of the limbs of salamanders and the tails of tadpoles. I got some tadpoles and severed their tails. Sure enough, there was absolutely no regeneration of the stump when minute amounts of beryllium were present in the water in which they swam. This oddball system appealed to me, but I did not think it would get me very far; it was not much better than rabbits. Then I hit upon plants. Simpler than tadpoles or rabbits, plants also contain chlorophyll, which is responsible for their color and for their ability to absorb light energy. Chlorophyll is an organic molecule in the middle of which sits an atom of magnesium,* similar to the way iron occupies a key site in the oxygen-carrying hemoglobin molecules in blood cells. Might beryllium compete with magnesium for its place in chlorophyll—substitute for it, perhaps—as the plant grew, thereby altering in some clearly discernible and measurable manner the plant's function? It was a long shot, but I was free to try it and my colleagues did not laugh when I talked about it at lunch.

I began growing tomato plants in bottles of simple salt solutions in a neighbor's greenhouse. (The MGH had no such facilities.) The plants were divided into four groups: those supplied with optimal magnesium for growth, some with and some without beryllium; and those given less than optimal magnesium, with and without beryllium. Beryllium had no discernible effect on the healthy growth of plants given optimal magnesium. Magnesium-deficient plants routinely sickened and died within ten days. But similarly

*To give some sense of relative sizes: Magnesium weighs 24 units, chlorophyll about 600, and an enzyme (a protein) about 100,000.

deficient plants, to my delight, flourished in solutions containing beryllium. Indeed, beryllium, a metal foreign to all biological systems, could replace about *half* of the plant's total magnesium requirement.

Convinced by experiments with hundreds of plants that the phenomenon was real, I proceeded to do a similar set of experiments in the laboratory with green algae (single-cell plants that could be grown in test tubes and whose growth could be even better quantitated) and obtained the same results. To nail down my hypothesis, I learned how chlorophyll could be extracted from plants and had an expert in spectroscopic analysis see whether any beryllium had gotten into the chlorophyll. It had not!

My failure to find beryllium enthroned in the center of the chlorophyll molecules was disappointing, but it did have the advantage of being unequivocal. And beryllium's stimulation of plant growth under conditions of magnesium deficiency remained a fact that needed explaining. Magnesium, after all, plays a critical role in many other cellular processes. I was impressed both with the results and with my ingenuity in having selected the model system. I had had a good testable—falsifiable—hypothesis and clearly proved it false.

Soon my assumption that I was the first to make such observations turned out to be false too. In digging in the literature preparatory to publishing the work, I found that at least two other scientists, back in 1888, had observed beryllium stimulation of plant growth and that one had even suggested the effect might be due to competition with magnesium. I also discovered, as I learned more chemistry, that the atomic size of beryllium (9 compared with magnesium's 24) precluded a good fit in chlorophyll. So, if I had known more, I would probably never have done the experiments! (I have known several scientists whose encyclopedic knowledge seemed to impede, mask, or substitute

Harriet Hardy, ca. 1950 *(courtesy Harriet Hardy)*

for whatever creative potential they might have had. Their ability to act—and science *is* action—was impeded by their erudition. They seemed to be weighed down by a sense that everything had already been done.)

I enthusiastically pursued the beryllium problem for two more years and independently published two papers showing that the metal did indeed compete with magnesium in influencing the activity of certain plant enzymes, as it did in animals.[3] Neither I nor anyone else, as far as I know, has further penetrated the powerful biological effect of this simple metal. It remains an intriguing mystery.

The beryllium project convinced me that I could generate a respectable idea, make use of a good model system, test the idea so as to produce a solid conclusion, and learn from it. I felt I had earned an operator's license to explore nature. During those three years, I became something of an expert on beryllium. I was invited to meetings, and my opinion was sought. I was indebted to Dr. Harriet Hardy, a protégée of Joe Aub's and an expert in occupational medicine with the Massachusetts Division of Occupational Hygiene and the Atomic Energy Commission. Her detective work had uncovered berylliosis among the fluorescent lamp workers, and she strongly encouraged me in my work. She was a warm, honest, enthusiastic physician who could be tough as nails with the fluorescent lamp manufacturers; she eventually forced them to abandon the use of beryllium and the practices that exposed their employees to high risk. She has, incidentally, written a lively account of her inspiring and fascinating career.[4]

I was pleased by my meteoric rise to fame but uneasy about how readily one could become an "expert." Indeed, it was this sense of having become superficially successful without much effort that impelled me to get more deeply involved in some solid biochemistry.

Half of the dozen scientists at the Huntington labs during this time were absorbed in a free-wheeling exploration of the mechanism of protein synthesis. This was an investigation, then in its embryonic stages, destined within a few years to illuminate the central question of biology: how did genetic information presumably stored in DNA get translated into the substance of life, most of which is protein? The group was headed by Paul Zamecnik, who more than any other single figure in science masterminded the attack on protein synthesis. He also, I am grateful to acknowledge, took an early interest in my development as a scientist. My beryllium project was peripheral to his group's project, but I found myself increasingly looking over its members' shoulders.

Every form of life, whether microbe, mosquito, mouse, or man, is an integrated and coordinated assembly of thousands of different types of protein molecules that form the structure of our cells and carry out their multitudinous functions—producing energy and consuming it, performing mechanical work, constructing cellular substance, transporting food and building units into, and waste materials out of, cells. The uniqueness of each living creature resides in the special structural and consequent functional characteristics of each of some fifty thousand different types of protein molecules. That is to say, hereditary information is expressed in the individual being as specific, functionally unique protein molecules.

Paul Zamecnik, ca. 1950

Proteins are large, complex three-dimensional molecules of many different sizes and shapes, each type of which is composed of thousands of atoms of carbon, nitrogen, oxygen, hydrogen, and sulfur. But every protein can be disentangled (denatured) into one or several straight chains of a few hundred amino acids each. The linkage between each amino acid in the chain is called a peptide bond; the chain is called a polypeptide. Exactly twenty amino acids are found in proteins in nature, and most proteins contain most of them. The amino acid composition of proteins and the order or sequence in which they are linked to each other in the polypeptide chain determine what shape, function, and other special characteristics that protein will have in its final configuration.

By 1950, Frederick Sanger in Cambridge, England, had laboriously worked out the amino acid sequence of the small protein insulin, which is only 120 amino acids in length. It was clear from this and all subsequent sequence analyses of proteins that there was no discernible plan or intelligible pattern in the amino acid order. In other words, the machinery that strung together amino acids into protein chains appeared to be instructed by chance events presumably emanating from a long evolutionary history.

In the late 1940s, while many of the ablest biochemists in the world had a general interest in the process of self-assembly and in how cells made proteins, it was not obvious how to go about getting a handle on the problem. To begin to understand how cells assemble protein molecules, one needed a way of unequivocally detecting newly made molecules among the enormously greater mass of protein molecules already present in cells. One way to do this would be to introduce into the system one amino acid that was tagged or labeled in such a way that it could be tracked as it became incorporated into protein molecules as they were being created.

Radioactive isotopes of important biological atoms became available after World War II as an offshoot of the atomic bomb project. Their use for labeling key body chemicals, such as amino acids, rapidly became biochemistry's most valuable tool for studying cellular dynamics. An amino acid whose normal carbon atoms (^{12}C) have been replaced by a radioactive isotope called carbon 14 (^{14}C) behaves in the body like the normal amino acid, but because it emits a radioactive signal, it can be followed wherever it goes. One can isolate cellular or body parts, separate out their proteins, and measure their radioactivity in a Geiger counter. The rate at which proteins acquire radioactivity measures the rate of new protein synthesis.

Robert Loftfield, a bright young organic chemist in Paul Zamecnik's lab, made the first amino acids labeled with ^{14}C soon after the war, giving the group a leg up in the exploration of protein synthesis. These amino acids were injected into rats; after an interval to allow the amino acids to pass from the circulation into the tissues, the animals were killed, their livers removed and ground up to break open the cells, the cell parts separated, the proteins extracted, and their radioactivity measured.

The group was soon getting solid data on rates of protein synthesis in different kinds of animal cells. To probe deeper into what individual cells were doing, the researchers began using thin slices of living tissue and incubating them directly with radioactive amino acids in test tubes. The cells in these slices actively made protein, too.

With the mildness and firmness that we all came to see as basic character traits of his, Paul Zamecnik insisted on performing extensive, difficult, and unexciting experiments to prove rigorously that bona fide protein synthesis was occurring in those systems. He needed to be sure that the radioactive amino acid had been integrally and properly linked into the polypeptide chain. After all, the amino acids might simply be sticking onto proteins in some artificial way, caused by the cell disruption process. This demonstration required breaking down proteins and showing that all radioactivity could be recovered as the originally labeled amino acid, which had been bound in proper peptide linkages to other amino acids in the chain. The system was then poised for further analysis.

In value as a research tool, the ultracentrifuge runs a close second to isotopically labeled compounds. The modern biological ultracentrifuge applies centrifugal forces several hundred thousand times the force of gravity to mixtures of biological materials, such as broken-up cells, for the purpose of separating the cell parts by their relative density. The ultracentrifuge made its most notable early contribution to protein synthesis in Zamecnik's and a few other laboratories by making it possible to show that as liver cells incorporate radioactive amino acids into their proteins, new protein is first found in the cellular organelles called ribosomes. These dense particles are sedimented to the bottom of a tube in the ultracentrifuge more rapidly than other cell components. Inside intact cells seen under powerful magnification in the electron microscope, ribosomes appear as specks peppered throughout the cytoplasm, the cell space outside the nucleus. They are rich in ribonucleic acid (RNA); indeed, it is in ribosomes that most of the RNA in cells can be found (a ribosome is a body rich in RNA). Earlier in the 1940s, Torbjörn Caspersson in Sweden and Jean Brachet in Belgium had observed that cells that tended to be fast growers contained more RNA, therefore more ribosomes, than slower-growing cells. Since rapid growth meant rapid protein synthesis, these were the first indications of a direct relationship between RNA and protein synthesis. The discovery that RNA-rich ribosomes were the actual sites of new protein manufacture nicely correlated with these earlier findings.

Another important early discovery, again made in Paul's laboratories, was that protein synthesis in incubated animal tissue requires oxygen. This was the first direct evidence that a cell supplies energy for the process of protein synthesis, that protein synthesis is somehow coupled to the energy-producing machinery in a cell. Such machinery, which makes chemical energy available by burning (oxidizing) food, was known to exist in cells but had not been shown to be connected functionally with protein synthesis. Paul found that a poison, known to prevent the production of energy from oxidation, stopped protein synthesis.[5]

I watched these beginnings in the study of protein mechanisms with growing fascination. What I learned in those first three years at the Huntington made me acutely aware of four things: how little I knew about biological mechanisms in general; how little anyone knew about the synthesis of large molecules; how ripe the field was for experimental exploration; and that if I wanted to be a serious investigator of living processes, I had better gain some more knowledge. I came to share the conviction of Paul and his associates that the way cells make protein was exciting as a career pursuit and could be a window on the control of growth in embryonic development and cancer, two of the most intriguing challenges in biological science. The beryllium project, on the other hand, seemed a bit like "applied" toxicological research, in which progress could be expected only if one knew a lot more about fundamental processes.

As I was winding up the first phase of my work at the Huntington, in 1951, science was awakening to a new way of looking at biology. Dramatic revelations were about to emerge—about the nature of the genetic material, its mode of replication, and the mechanism by which it translated its information into living substance. The broad outlines of the field I was about to enter at the beginning of the 1950s could be summarized as follows.

The assembly of a living being is an intricate process of creating order—of putting randomly distributed small building blocks into nonrandom, very long, functional sets, or polymers. The molecules involved in this process could be seen as participating in tiny realms of the construction of living orderliness within a vast sea of disarray. To accomplish this reduction of disorder, the living ordering system required (1) a *plan* of the order to be achieved; (2) the *building materials*; (3) the *energy* or power to accomplish the task; (4) the *machinery* to assemble the finished product, including a means of funneling energy into the process; and (5) the ability to *reproduce* efficiently and accurately the first four requirements so that the process of building a new organism could be repeated generation after generation. A sixth item, the ability to *change* and therefore to *evolve,* was essential to provide the variety of life forms that exist and have ever existed.

In place in the United States, Great Britain, France, and Japan were two entirely different intellectual, methodological, and temperamental approaches to tackling the life assembly problem. One, which attracted scientists comfortable with the ways of physics and genetics, focused mainly on *the plan*—in particular, the elucidation of the structures of very big molecules. It concentrated on these questions: How is genetic information encoded, and how is it expressed? What molecular principles govern the accurate replication of genetic information from generation to generation? This alignment of genetics and physics and chemistry to focus on the relation of information to molecular structure was the intellectual force that brought the term *molecular biology* into vogue in the 1950s.

The other approach, that of more traditional biochemistry, to which I found myself gravitating, sought, by breaking open cells and analyzing their parts and enzymatic activities, to reveal the machinery that carried out the process of protein synthesis. The energetics of the process and the detailed chemical linking of single units (monomers) to make polymers were the chief points of focus. At the beginning of the 1950s, neither camp had much awareness of or interest in the other, but at the end of the decade they would become companions in common purpose.

By the early 1950s, genetics had shown that discrete cellular structures and functions in the form of proteins were specified by discrete entities called genes. Genes resided in chromosomes, which in turn were localized in the cell nucleus. Chromosomes contained deoxyribonucleic acid (DNA) and protein, and the former had been tentatively identified as the material of inheritance. In other words, genes were made of DNA.

Back in the 1920s, Frederick Griffith in England had made the startling discovery that material extracted from killed bacteria could permanently alter (transform) the inheritance of living bacteria. In 1944, Ostwald Avery, Colin MacLeod, and Maclyn McCarty drew the pivotal conclusion that the active transforming component in this material was DNA. (Published in 1944, this momentous curtain raiser of the so-called biological revolution is described by McCarty in the first volume of this series.) While the theory of bacterial transformation by DNA was impressively documented, it was not fully accepted by scientists, partly because the composition of DNA was much simpler than that of protein. Proteins, which come in myriad sizes and shapes and are made of a larger variety of units, were favored as candidates for the genetic material. It was not until nine years later, when Watson and Crick revealed the remarkable properties inherent in its structure, that DNA received universal acceptance as the material basis of inheritance.

DNA certainly had all the attributes of the *plan.* It

appeared to be the only temporal and chemical link between generations—that is, it was observably the sole material contribution that parents made to their offspring. It is an immensely long chain made up of four different links or units called nucleotides (adenylic, guanylic, cytidylic, and thymidylic acid). Each nucleotide is a ringlike structure composed of carbon and nitrogen atoms (the base) linked to a sugar molecule (deoxyribose), which in turn is linked to a phosphorus-oxygen group (phosphate). A single DNA chain is a repeating sugar-phosphate backbone from which the bases protrude at right angles. In nature, it is a double chain in which the protruding bases form weak linkages with each other according to their shapes: adenylic acid always pairs with thymidylic acid, guanylic with cytidylic. A chain of DNA could thus be viewed as a one-dimensional chemical text, written in a four-letter language, comprising instructions for the assembly of a living organism.

Another critical set of discoveries in the 1940s was that bacteria, the smallest, most abundant free-living creatures on Earth, carried out intracellular processes much the same way animal cells did and that they were as subject as animals were to the guiding evolutionary forces of mutation and natural selection. These discoveries kicked off the widespread use of bacteria—and the viruses that preyed upon them—as experimental models, a development that was to be astonishingly fruitful in the coming decade.

Finally, by the start of the 1950s, it was known through the work of George Beadle and Edward Tatum on the fungus *Neurospora*, and through subsequent work on bacteria, that there was a one-to-one relation between a gene and a single polypeptide chain. That is to say, each of our genes is responsible for the production of one particular polypeptide. (The polypeptide, once made, then folds into a complex, functional, three-dimensional protein.)

The grand hypothesis of that decade was that heritable

information, residing in the linear sequence of the bases in an organism's DNA, was replicated at each cell generation and passed to progeny, where the base sequences were translated into sequences of amino acids in protein molecules. One gene, a segment of several hundred DNA bases, "coded for" one polypeptide chain. The conversion of genetic information into living substance was thus a simple language translation process. In a landmark paper, Francis Crick, biology's theoretician par excellence, formulated these ideas for those in the field as the *sequence hypothesis.* This publication also promulgated what Crick called the *central dogma,* which simply and plainly declared that once information had passed into protein, it could not get out again.*

The beauty of these hypotheses was that they were testable. Among the specific questions to be asked were the following: What is the actual physical structure of DNA in which information is stored and which is capable of reproducing itself? What is the cipher or code—that is, what set of DNA bases specifies each of the twenty amino acids in protein? Is the code in fact laid out linearly, in parallel with the sequence of amino acids in protein? What is the actual mechanical process used to achieve the building of a protein molecule?

In 1951, the last question relating to the *machinery* for making protein—that is, for "reading" DNA instructions into protein substance—was just beginning to reveal the shad-

---

* For his book *The Eighth Day of Creation* (New York: Simon & Schuster, 1979), a remarkable account of the development of molecular biology, Horace Judson asked Crick why he had called this idea the central *dogma.* Crick replied, "Ah! That's a very, very interesting thing! It was because, I think, of my curious religious upbringing. . . . Because Jacques [Monod] has since told me that a dogma is something which a true believer *cannot doubt!* . . . But that *wasn't* what was in *my* mind. [I thought of dogma as an] idea for which there was *no reasonable evidence.* You see?!" [And Crick gave a roar of delight.] "I just didn't know what dogma *meant.* And I could just as well have called it the 'Central Hypothesis,' or—you know. Which is what I meant to say. Dogma was just a catch phrase." (p. 337)

ows of an answer. Ribosomes had been identified in Paul Zamecnik's and other laboratories as the likely site of protein synthesis, and bits of evidence pointed to an energy requirement for the process, but little else was known about what proved to be a beautifully intricate process.

The laws of physics and chemistry do not allow the making of order from chaos without energy. If proteins are to be constructed, energy has to come from somewhere. Paul Zamecnik was profoundly influenced by the powerful intellect of Fritz Lipmann, who worked in neighboring labs at the MGH and had become identified with issues of biochemical energetics. Lipmann had published an influential review article that mustered evidence to suggest that the potential energy residing in bonds involving phosphate groups (phosphorus-oxygen molecules) was the immediate power supply for protein synthesis. The dissection of cellular mechanisms of protein synthesis was right at the cutting edge of biological science, and I wanted to be there with it. With Paul's assurance that I could have a place in his group upon my return, I decided to apply to the American Cancer Society for a fellowship that would allow me to take two years away from the Huntington to "retool"—particularly to learn more about protein chemistry and biochemical energetics.

4

✤ ✤ ✤

Amino Acid Activation

Earlier in his career, Paul Zamecnik had worked with the noted physical chemist Kaj Linderstrøm-Lang at the Carlsberg Laboratory in Denmark. Paul had chosen an inauspicious time for his European sojourn, however. He and his wife, Mary, arrived in Copenhagen on September 1, 1939, the day Nazi Germany invaded Poland. They got out by the skin of their teeth shortly after April 9, 1940, the day the German troops occupied Denmark. Paul remembered Copenhagen and Linderstrøm-Lang with great fondness, however, and suggested that I spend a year at the Carlsberg to learn more about protein chemistry. I got my fellowship from the American Cancer Society and, in June of 1951, settled in Copenhagen with my wife and three children for fifteen months.

The Carlsberg Laboratory story is remarkable. It was founded by J. C. Jacobsen, the son of a successful brewer in Copenhagen. A talented man who showed an early interest in science, he worked in his father's brewery and took over its direction upon his father's death, in 1835. Danish beer had little to recommend it at the time, and Jacobsen, determined to improve it, traveled to Munich to study the methods of the Germans; he returned to Copenhagen "carrying, with great difficulty, a tin of genuine Bavarian yeast in his hat-box."[1] In 1847, he built a new brewery in Copenhagen, the now famous Carlsberg Brewery.

Fermentation was a tricky process to control for Jacobsen and indeed for the whole industry until Louis Pasteur solved most of the problems in the 1860s. Jacobsen was fascinated by Pasteur's research and set up a brewing laboratory in 1871. Then, in 1875, he established the Carlsberg Foundation to promote scientific research. He endowed it in perpetuity with funds from the brewery and asked the Royal Danish Academy of Science and Letters to appoint trustees from its membership to supervise and guide its scientific activities. In his request to the academy, he stated:

> Since such an institute, designed for specialized studies, can only thrive if it is supported by the spirit of science and is pervaded by the light which emanates from the sciences as a whole, and since this light has for me been a source of joy and well-being, it is of vital concern to me, in order to repay some of my indebtedness, to make a contribution to the general advancement of science, especially in those directions where it appears to me the state has not been able to afford all the necessary resources and is unlikely to do so in the future.[2]

Jacobsen's wisdom and generosity enormously enhanced his small country's reputation as a notable center of first-rate science. As a small institute with a permanent staff of some dozen scientists and laboratory assistants, and with a steady flow of visiting scientists, the Carlsberg Laboratory has had a record of exceptional achievement, including the training of scores of scientists, many of whom have gone on to positions of eminence in other countries. It has encouraged open and honest inquiry in a relaxed, noncompetitive environment.

Kaj Linderstrøm-Lang, known to all as Lang, was head of the Carlsberg's chemical division and a major figure in world science. He pioneered the use of micromethods in biological chemistry and made brilliant investigations of the physicochemical properties of proteins. He also was a gifted mathematician, musician, painter, and raconteur. He delighted audiences, all too rarely, with a hilarious scientific seminar entitled "The Thermodynamic Activity of the Male Housefly (*Muscus domesticus* L.L.)." The work was published, coauthored by one F. Fizz-Loony, in 1956 by Academic Press, and widely circulated among his friends. Lang applied his skill as a caricaturist to the laboratory's visiting scientists. Often bitingly insightful, these drawings were once a year displayed as a rogues' gallery in the building's entrance hall. Lang also manufactured his own fireworks. His personal celebration of the Fourth of July from the front lawn of his home, which adjoined the Carlsberg Laboratory building, was enjoyed not only by his visiting guests, most of whom were American, but by the whole city of Copenhagen as well. Blue eyed, sardonic, always accompanied by a large fuming cigar, Lang was a man of sensibility, warmth, and great humanity. He was honored by his countrymen as a hero of the Danish underground for having helped many Jews in their escape from Germany via Denmark to neutral Sweden during World War II.

Another luminary of the Danish scientific community whom I got to know that year was Herman Kalckar, a gifted professor of biochemistry at the University of Copenhagen who had contributed prodigiously to knowledge of cellular synthetic processes and their regulation. He was fifteen years my senior—then a distant hero of biochemistry whose work I had studied and admired before I went to Denmark. Our friendship, nurtured in the easy atmosphere of Copenhagen, showed me for the first time the relative unimportance of age, rank, and prestige in the relationships of scientists.

Kaj Linderstrom-Lang, 1952 *(Carlsberg Laboratory archives)*

Herman Kalckar, ca. 1960 *(courtesy Herman Kalckar)*

I cannot say that the Carlsberg sojourn contributed greatly to my technical training as a biochemist, but the laboratory's serious dedication to and enthusiasm for precise physicochemical analysis, together with its European urbanity and unhurried scholarly dedication to the search for explanation, certainly left something with me. The experience confirmed my growing conviction that, as an environment for research, the institute was superior to the traditional university department. With Lang's approval, I worked on an idea of my own that did not pan out, talked a lot, and drank a great deal of free Carlsberg beer during discussions on science, politics, and the many fascinating differences between the Danish and the American cultures. It was the time of Joseph McCarthy's shameful performance in the United States: the Danes were appalled and I was embarrassed, as though it were somehow my fault. I immensely enjoyed the many attractions of that delightful and civilized country. Copenhagen was a leisurely, gracious, storybook city. My memory of it is associated with bicycles, thousands of them. Their riders, particularly the women with billowing skirts and bare legs, gave the city an air of lightness and beauty and youth and health. I formed a number of enduring friendships there. The richest was that with Paul Plesner, with whom I have shared research interests and good conversation over the years since. He was then a physician-researcher just getting his start, as I was at the Carlsberg, and is now a professor of molecular biology at Odense University in Denmark. I watched with delight as my children became proficient in a new language, and I wept as our ship pulled out of Copenhagen's harbor on the return voyage to America.

Linderstrøm-Lang, Kalckar, and Lipmann independently all had the vision to see that energy for the linking of amino acids must be funneled into the process from some other cellular energy-producing event. It is hard to believe now, but most of the thinking about protein biosynthesis before this time had favored the idea that proteins were made by a reversal of the enzymatic process by which they were broken down in cells—that is, by a reversal of proteolysis. These three men and Paul Zamecnik very much conditioned my thinking and influenced the subsequent course of my research.

Paul Plesner, 1985 *(courtesy Paul Plesner)*

Fritz Lipmann, 1953 *(Massachusetts General Hospital archives)*

Before leaving for the Carlsberg, I had arranged that my second year away from the Huntington be with Fritz Lipmann back at the MGH's Biochemical Research Laboratory. So, in September 1952, I plunged into that boiling pot of cell metabolism and biochemical energetics. After the leisurely scientific pace of Europe, in which I felt very much at home, I found the seething enthusiasm and competitiveness in Lipmann's lab a bit daunting at first. But I soon became familiar with the concepts and techniques that would within a year and a half lead me to the discovery of the first step in protein synthesis—the mechanism by which amino acids are energized or activated.

Fritz Lipmann was an unpretentious, shy, awkward, gentle man with a wide-ranging scientific vision. He ran his lab of a dozen scientists, postdocs, and assistants in a deceptively casual manner. Lipmann's vague, absent-minded-professor style generated innumerable stories that have delighted his colleagues over the years.

I remember when he gave a lecture in a large hall at Harvard, speaking hesitatingly, shuffling back and forth behind a long chemical bench. As he spoke, he inserted an index finger into a hole in the bench and to his and then the audience's consternation, could not get it out. Several minutes passed as the audience, at first with silent sympathy and then with increasing amusement, watched Lipmann struggle to extricate his finger and try to continue his lecture at the same time. Finally, to a burst of applause, the abused finger popped free.

I was impressed by the departmental seminars, which were stamped with his personality: fifteen or twenty scientists from the lab and other hospital departments, Lipmann presiding at the head of the table, quiet, seemingly preoccupied. The speaker finishes—a long silence—one minute, two, three—awkward—I try to think of a question—then Lipmann, looking down at the table or perhaps out the window, asks *the* incisive, perceptive, exactly right first question. The dam breaks, and a lively discussion ensues.

After training in medicine in Germany, Lipmann decided he did not want to go into practice. "An important motive," he said, "was my uneasiness about eventually accepting its commercial aspects: to buy one's health and accept individual payment for supplying it."[3] He gravitated instead toward biochemistry. After retooling in chemistry, he spent three years with the great Otto Meyerhof in Berlin and Heidelberg, where he was launched into his life's interest in energy utilization in muscular and biosynthetic work. For Lipmann, too, there was a Copenhagen connection. He spent seven of his most productive years there, from 1932 to 1939, in the laboratory of Albert Fischer. In those years, Lipmann got to know and admire Linderstrøm-Lang, who helped him when the Nazis began to menace Denmark.

Lipmann came first to Cornell Medical School in 1939 and then to Boston in 1941. Among those responsible for bringing him to MGH's Department of Surgery was my surgeon hero of medical school, Oliver Cope. (Cope also played a critical role in Paul Zamecnik's career. Paul had applied for an internship in surgery at MGH and was put on the waiting list. Cope suggested to Paul that he see Joseph Aub, who might have a residency in medicine available that Paul could fill in the interim. Paul did so and got the residency, which in turn opened the door to the Huntington for him a few years later.)

In that same year, Lipmann's insightful and prophetic review article on the metabolic function of phosphate-bond energy was published.[4] In Copenhagen, Lipmann had discovered that a compound called acetyl phosphate (acetic acid linked to a phosphate group) was produced in bacteria when they oxidized carbohydrates. Acetyl phosphate is an important form of construction energy because it can donate its acetate part to other molecules in the process of making, for example, larger polymers of acetate—that is, fat molecules. The acetate is energized, made capable of linking to other molecules, by virtue of its attachment to phosphate. The bond between the acetyl group and the phosphate group, then, can be said to be energy rich—that is, if it is split apart, a lot of energy is released as heat. Lipmann therefore called acetyl phosphate active acetate. Acetyl phosphate turned out to be important only in bacteria, but it led Lipmann to his insights on the general importance of phosphate-bond energy in cellular construction. If acetate could be energized—activated—by phosphate, why couldn't other molecules be similarly energized? Acetyl phosphate led Lipmann to the discovery of acetyl–coenzyme A (CoA), which turned out to be a more important general form of active acetate in all living tissues. (In bacteria, acetyl phosphate was converted to acetyl-CoA by a specific enzyme.) Lipmann isolated and chemically characterized CoA. For this set of discoveries, he was awarded the Nobel Prize in 1953—a grand and happy occasion at the MGH for all of us working there at the time.

The principal focus in Lipmann's lab during 1952 and 1953 was trying to understand the details of how acetate got attached to CoA—in other words, how the activation of acetate was accomplished. Acetate—acetic acid—is as simple as it is important. Its chemical structure may be written thus:

```
     H
     |
H—C—COOH
     |
     H
```

After Lipmann's discovery of CoA, others had found that the coenzyme contained a sulfur atom and that acetate was linked to CoA by a bond between acetate's carboxyl group

(COOH) and the sulfur atom (S) on CoA, losing a hydrogen and oxygen atom in the process. The resulting compound,

```
     H
     |
H—C—CO—S—CoA
     |
     H
```

was the activated form of acetate found in all cells.

Lipmann knew that the formation of active acetate, acetyl-CoA, required energy from the phosphate bonds of another cell chemical, adenosine triphosphate (ATP). How this process actually occurred chemically was a subject of intense investigation in his and other labs in the United States and in Europe. It was a central problem because its solution looked to be the likely entryway to a general understanding of how energy got funneled into synthetic processes. Lipmann in his 1941 review article on energy transfer mechanisms[5] had speculated, in an analogy with acetate activation, that amino acids, which also have a COOH group, might similarly be activated by interaction with the phosphate bonds of ATP.

Let me pause here to introduce ATP more fully. ATP is the basic form of chemical energy in all cells. When carbon compounds, as food, are oxidized, much of the energy produced is chemically built into ATP molecules by a complex oxidative process that takes place in cell bodies called mitochondria. The base adenine linked to the sugar ribose is called adenosine, which I here abbreviate Ad. Adenosine linked to one phosphate group is adenylic acid, Ad · P; linked to two additional phosphates, it becomes ATP, Ad · P ~ P ~ P. The bonds between the two outer phosphates, indicated by the mark ~, are the ones in which energy is "stored"—that is, they represent a pool of potential, usable energy, as does the bond between acetate and phosphate in acetyl phosphate. In the cell, most of these bonds supply energy to do chemical work: for transporting molecules into and out of cells, for activating muscle fibers and thus making motion possible, and for synthesizing all body constituents—notably proteins. *How* this process of funneling energy into biosynthesis occurred was the big mystery.

Amino acids contain a carboxyl group like acetate. The structure of all twenty amino acids can be written by a general formula thus:

$$\begin{array}{c} \mathrm{H} \\ | \\ \mathrm{R{-}C{-}COOH} \\ | \\ \mathrm{NH_2} \end{array}$$

where R is a different chemical group for each amino acid. Once activated, all by the same mechanism, the amino acids were presumed to be ready to enter into the cell machinery for linking them in chains in their proper order.

What happened next nicely illustrates the vagaries of fortune in science. The discovery of the mechanism of amino acid activation, the presumed first step in protein synthesis, had become one of the most sought-after goals of biochemistry. Lipmann had spent more time than anyone else pondering the problem and had given the scientific community an analogy (the activation of acetate) and a hypothesis (activation by use of the phosphate bonds of ATP) that could be tested—although *how* was not then clear.

For years, Lipmann had used a simple model for detecting acetate activation in crude extracts of tissue. He would simply add some sulfanilamide, an acetate acceptor, to a reaction in which he wished to detect acetate activation. Activated acetate would readily link to the sulfanilamide and cause an easily detectable color change. This linkage between the acetate and sulfanilamide is between the COOH group of the acetate and an amino group (—NH_2) on the sulfanilamide. It is, in fact, a peptide bond (CO—NH), exactly the same as that between amino acids in a peptide chain:

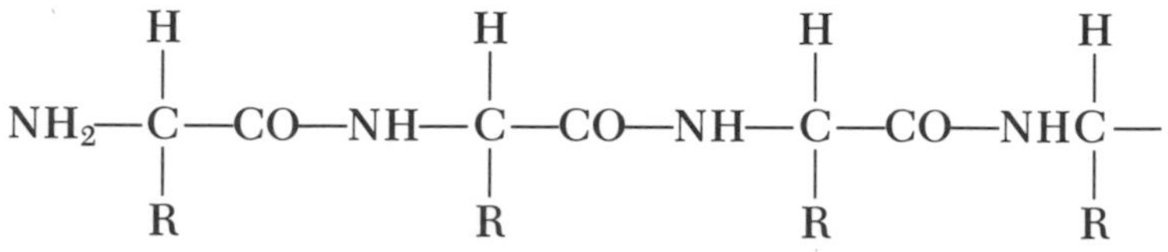

$$
\begin{array}{ccccccc}
 & H & & H & & H & & H \\
 & | & & | & & | & & | \\
NH_2— & C & —CO—NH— & C & —CO—NH— & C & —CO—NH & C— \\
 & | & & | & & | & & | \\
 & R & & R & & R & & R
\end{array}
$$

It was this similarity between the acetyl-sulfanilamide bond and the true peptide bonds of protein that had led Lipmann to suspect that very similar mechanisms must be involved in both processes. (It should be said, too, that other investigators in the late 1940s were beginning to get evidence as to how other model peptide bonds were synthesized.[6])

In 1953, however, Lipmann's experimental goal had not deviated from the details of acetate activation; he had not attacked amino acid activation. During my year in his lab, Lipmann and his colleagues did experiments and published a paper on the mechanism by which ATP might activate acetate, which turned out to be wrong. Paul Berg of Washington University, in St. Louis, forthwith published the correct mechanism and so scooped Lipmann.[7]

While all this was going on, I had been involved in a project investigating how cells make CoA. Three of the steps in the synthesis of CoA, starting with the B vitamin pantothenic acid, require ATP. David Novelli, a longtime associate of Lipmann's, and I worked out one of the steps by which CoA is synthesized.[8] While this work familiarized me with some valuable techniques and concepts, it had little direct influence otherwise on the discovery of amino acid activation. However, a technique being used by another visiting scientist in the lab, Werner Maas, had a more immediate bearing on the problem.

Bacteria need the vitamin pantothenic acid to make CoA, but being more versatile than we, they can make the vitamin themselves (so for them it is not a vitamin). Maas was studying how bacteria accomplish this. He found that they had to activate a molecule called pantoic acid and then link it to a special amino acid β-alanine. And they used ATP for the purpose.[9] This linkage was, again, a peptide bond, identical with that between amino acids in protein. To follow this activation process, he used a technique called phosphate-ATP exchange.

Ordinarily when one studies a chemical reaction, one puts in a measured amount of reactant, allows the reaction to proceed, and then measures the amount of reactant used up or the amount of product formed. But, in searching for reactions in which very small amounts of substances are being converted, there may not be enough product to measure. In these circumstances, exchange reactions involving radioisotopes can be of great value, for they are highly sensitive and can detect very small amounts of chemical reaction. This technique, described below, proved to be a critical lead for me when I began to search for evidence of amino acid activation.

So there I was in the hotbed of biochemical energetics, working under the wing of the man who had made the most prescient guesses about how amino acids were energized for protein synthesis, and no one was working on the problem![10] As I watched all this going on, I was itching to get back to the Huntington and protein synthesis and to put to use the techniques I was learning.

The timing of my return to the Huntington in 1953 was propitious, not only because of the developments in Lipmann's lab but also because of further progress Paul and his colleagues were making in dissecting out the liver protein–synthesizing machinery. After his success in getting liver cells, in the form of thin slices of tissue, to make protein, the logical next step was to get *parts* of broken-open cells to make protein so that one could see which components were required for the process. About the time I was leaving for Copenhagen, Dr. Philip Siekevitz, a visiting scientist working in Paul's lab, was successful in getting some protein synthesis to occur in a dead, broken-cell preparation. Soon afterward, a new technique of gently breaking cells was developed by another researcher in the Huntington group, Nancy Bucher, who was then successfully coaxing broken-cell preparations to make cholesterol. Using her procedure, one could get respectable amounts of protein synthesis and show that it proceeded on ribosomes only in the presence of ATP.

Paul and his associates, Dr. Betty Keller and Dr. John Littlefield, were soon able to delineate the protein-making machinery of liver cells and of certain tumor cells as consisting of the following key components: (1) a crude "soluble" fraction, that is, the cellular material that remained in solution after the ribosomes had been centrifuged down, consisting mostly of protein but with a small amount of associated RNA; (2) the ribosomes upon which the new protein molecules were assembled; (3) amino acids; (4) ATP; and (5) *another* compound like ATP, GTP (guanosine triphosphate). GTP had been found by Betty Keller and Paul to be an additional requirement for protein synthesis. We would learn that this energy-producing compound was needed at a later step in protein synthesis, on the ribosome.

Paul agreed that I should tackle amino acid activation,

and I eagerly got at it. On the basis of the experience with acetic acid and pantoic acid activation, it seemed likely that the carboxyl group of an amino acid would link up with one phosphate of ATP, creating a high-energy (~) bond between amino acid and part of ATP. This would have the effect of preserving the chemical energy in a (~) bond, *conferring it upon* the amino acid.

Since Ad · P ~ P ~ P had two reactive bonds (~) and since a carboxyl group could react with either side of the bond, there were four theoretically possible reactions:

1. Carboxyl links up to Ad·P of Ad·P~P~P and ejects the rest of the molecule, P~P (called pyrophosphate, two phosphates linked together):

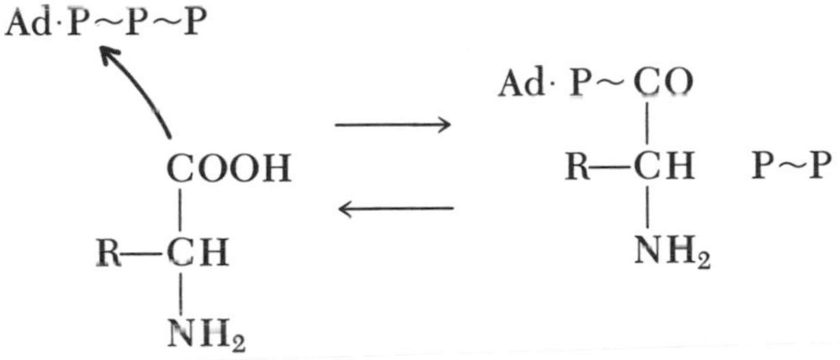

$$\begin{array}{c} \text{Ad}\cdot\text{P}\sim\text{P}\sim\text{P} \\ \\ \text{COOH} \\ | \\ \text{R—CH} \\ | \\ \text{NH}_2 \end{array} \quad \begin{array}{c} \longrightarrow \\ \longleftarrow \end{array} \quad \begin{array}{c} \text{Ad}\cdot\text{P}\sim\text{CO} \\ | \\ \text{R—CH} \\ | \\ \text{NH}_2 \end{array} \quad \text{P}\sim\text{P}$$

2. Carboxyl links up to Ad·P~P and ejects P:

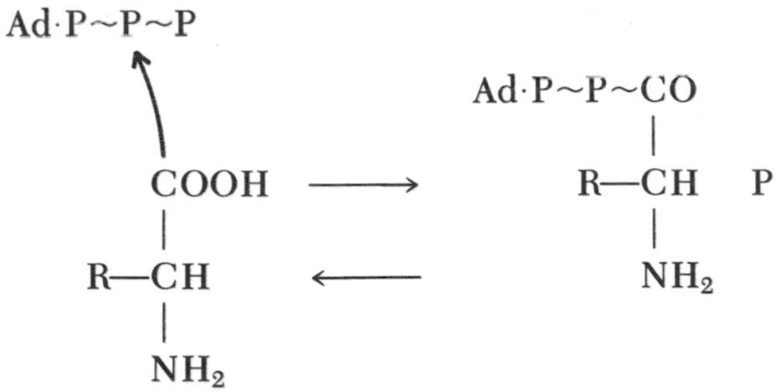

$$\begin{array}{c} \text{Ad}\cdot\text{P}\sim\text{P}\sim\text{P} \\ \\ \text{COOH} \\ | \\ \text{R—CH} \\ | \\ \text{NH}_2 \end{array} \quad \begin{array}{c} \longrightarrow \\ \longleftarrow \end{array} \quad \begin{array}{c} \text{Ad}\cdot\text{P}\sim\text{P}\sim\text{CO} \\ | \\ \text{R—CH} \\ | \\ \text{NH}_2 \end{array} \quad \text{P}$$

3. Carboxyl links up to P~P and ejects Ad·P:

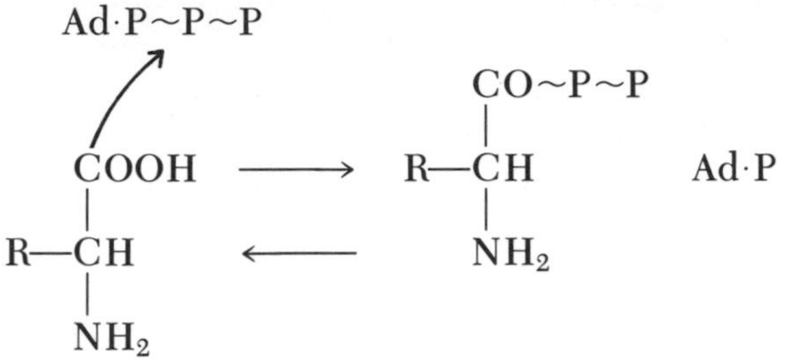

$$
\begin{array}{ccccc}
\text{Ad}\cdot\text{P}\sim\text{P}\sim\text{P} & & & & \\
\nearrow & & \text{CO}\sim\text{P}\sim\text{P} & & \\
 & & | & & \\
\text{COOH} & \longrightarrow & \text{R—CH} & & \text{Ad}\cdot\text{P} \\
| & & | & & \\
\text{R—CH} & \longleftarrow & \text{NH}_2 & & \\
| & & & & \\
\text{NH}_2 & & & &
\end{array}
$$

4. Carboxyl links up to P and ejects Ad·P~P:

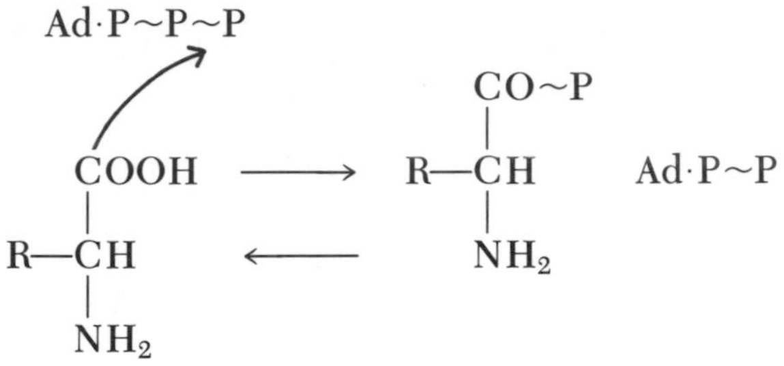

$$
\begin{array}{ccccc}
\text{Ad}\cdot\text{P}\sim\text{P}\sim\text{P} & & & & \\
\nearrow & & \text{CO}\sim\text{P} & & \\
 & & | & & \\
\text{COOH} & \longrightarrow & \text{R—CH} & & \text{Ad}\cdot\text{P}\sim\text{P} \\
| & & | & & \\
\text{R—CH} & \longleftarrow & \text{NH}_2 & & \\
| & & & & \\
\text{NH}_2 & & & &
\end{array}
$$

In each of the four cases, the carboxyl group of the amino acid pushes out and takes the place of a part of the ATP molecule. In this way, the energy in the substituted bond, instead of being released as useless heat, is preserved in the new chemical linkage between COOH and P, and the carboxyl group (and thus the amino acid) is said to be activated.

Since the energy in ATP on one side of the equation is equivalent to the energy in the "activated" amino acid on the other side of the equation, basic chemical principles dictate that the reaction will go readily (if enzymes are present in the cells to catalyze the reaction) in either direction—indicated by the double arrows. The direction favored at any instant depends on the relative amounts of the reacting substances on either side of the equation. If there is a lot of ATP and amino acids, the reaction will go faster to the right. If there is more activated complex and ejected product (Ad · P ~ amino acid and P ~ P in the case of reaction 1), it will go faster to the left. If the two sides balance, the reaction is said to be in equilibrium.

This reversibility was what allowed me to discover the secret of amino acid activation by using phosphate-ATP exchange. If you put a bit of a radioactively labeled compound into a system that is carrying out this kind of equilibrium reaction—for instance, into the right-hand side of the reaction—it will find its way back into the molecule whence it came on the left. For example, if I have some cell material that is able to perform reaction 1 and I add some radioactive P ~ P, it will mix indiscriminately with the nonradioactive P ~ P that is being produced in the reaction and find its way back into ATP by way of the reverse reaction—the reaction proceeding to the left. *The ATP phosphates will then become radioactive.* By reisolating the ATP after a period of incubation, I could measure its radioactivity.

For various reasons, reaction 2 seemed at the time the most likely and was the one I tried first. I took the soluble cell fraction that Paul had recently shown to be essential for protein synthesis, added some ATP and some radioactive phosphate (P) and a mixture of amino acids, and looked for radioactive ATP. None appeared. Nice clean result. Then I got some radioactive pyrophosphate (P ~ P) and set up reaction 1. (P ~ P is rapidly broken down by tissue enzymes, so I also added some fluoride, which prevents that breakdown.) It worked: ATP rapidly became radioactive. If I left out the amino acids, only a small amount of labeled P ~ P entered ATP—probably because of traces of amino acids in the soluble cell material. The more amino acids I added, the more radioactivity entered ATP. It was as simple as that.

(The cell material that performed this exchange reaction

indicative of activation was the so-called soluble fraction referred to above—material that remained in solution after the ribosomes had been centrifuged down in an ultracentrifuge. This soluble material was very crude but had been mostly freed of unwanted small molecules—particularly ATP and amino acids.)

Here, then, was strong evidence that the energy input event inaugurating the synthesis of a protein molecule was an attachment of the adenylic acid (Ad · P) portion of ATP to the carboxyl group of the amino acid. We pictured this activation reaction as follows:[11]

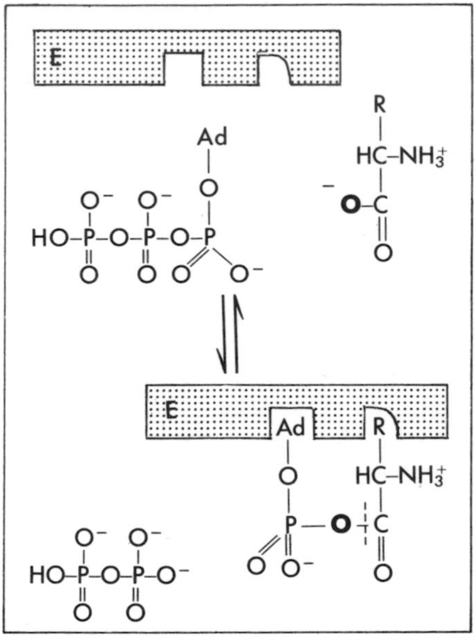

All reactions in cells are catalyzed—that is, promoted or made to proceed at acceptable rates—by enzymes. We assumed that an enzyme (labeled above as E), a protein molecule with a special affinity for a particular amino acid and ATP, bound these molecules and thereby promoted the formation of the activated intermediate (Ad · P ~ amino acid) on the enzyme surface and at the same time ejected the product P ~ P. This system remains in equilibrium, shuttling ATP and amino acid onto and off the enzyme. It is this reaction that is detected by radioactive P ~ P exchange with ATP. No measurable amounts of products accumulate, because the enzyme, like all catalysts, exists in minute quantity and the activated adenyl-amino acid is firmly bound to the enzyme. There is, in the system, nothing for the activated amino acid to react *with*. The reaction can be made to accumulate measurable amounts of product (that is, in the figure, to proceed downward) by adding a compound to which the amino acid will become irreversibly attached. A convenient chemical acceptor of activated amino acids is the simple molecule hydroxylamine. It reacts with the amino acid to form a hydroxamic acid, freeing the enzyme to pick up more ATP and amino acid and form more product. When I added hydroxylamine to the system, hydroxamic acids of each of the amino acids were formed, and P ~ P emerged in equal amounts. That clinched the correctness of the postulated mechanism.

Jubilation was high in the Huntington camp, and other colleagues were hot on the same trail. Paul Berg, having gotten the acetate activation reaction right, was going after amino acid activation by using exchange reactions. So was David Novelli in Cleveland, where he had moved after leaving Lipmann's lab. The experience brought home to me that the excitement of discovery lay not so much in doing something unique, which others would have done soon if I had not, but in being the *first* to do it. Seeing something that no one had seen before, but for which many were looking, was enormously exhilarating.

I proceeded to do some additional experiments and to write up the work for publication. It appeared in early 1955.[12]

A more definitive paper was published a year later, in January 1956.[13] In between, Berg and Novelli published their evidence for amino acid activation. We were all in agreement.

5

✤ ✤ ✤

Transfer RNA

Scientists are communicators par excellence; the process is their life's blood. They freely circulate written accounts of their investigation among their colleagues, both before and after publication; they run up astronomical telephone bills in discussing their work; they peruse one another's grant applications; they incessantly visit one another's laboratories, giving lectures, talking, and doing experiments using one another's equipment. They attend innumerable scientific meetings, and worry to varying degrees that their colleagues might steal their ideas or do their experiments before they can. On the whole, the habit of candor and the knowledge that science thrives on it keep the system open. And it is all global in scope, language differences never being a significant hindrance to understanding. I once spent a couple of hours with a Japanese colleague discussing details of experiments on protein synthesis—waving about numbers, graphs, diagrams, sketches, and hands in lieu of a knowledge of one another's language. We were both a bit tired at the end, but we understood each other.

There are, of course, fraternities, cliques, and clubs—domains of common interest and tradition, parochialism, and disdain for the outsider. Within these loose enclaves, formal lines of communication and grapevines work well. Between them, they may work poorly.

As I indicated earlier, in the 1950s and early 1960s a wide gulf separated scientists who focused mainly on molecular structure, information, and coding from scientists who were grappling with the mechanism of protein synthesis. The former used bacteria and viruses as experimental models. They were often contemptuous of the biochemists, who focused on the dissection of cellular machinery, most often in animal cells. The biochemists tended to view the molecular biologists as party crashers, whose novel use of physics and genetics was glamorously overshadowing the more traditional biochemical approaches. I remember the chagrin I felt when I presented my discovery of amino acid activation in Detroit in 1955 at a symposium organized by microbiologists. Presiding were Bernard Davis of Harvard (soon to be my boss) and Jacques Monod of the Pasteur Institute. I was the only participant working on animal cells. My lecture and the following brief discussion proceeded in an atmosphere of imperious disdain from the podium, as though I had entered the Ritz dining room without a tie. (That experience, however, was dwarfed by that of my colleague Marshall Nirenberg when he reported his momentous, Nobel Prize–winning experiments at the Fifth International Congress of Biochemistry, in Moscow, in 1961, that opened the door to the cracking of the genetic code. As a biochemist from the National Institutes of Health, unknown or ignored by the molecular biologists attending the meetings, Nirenberg spoke to a nearly empty room. Fortunately, shortly after his talk, Francis Crick, who recognized the importance of the work, arranged for Nirenberg to present his work again, to a substantial audience.[1])

In general during the 1950s, while the molecular biologists looked down upon the traditional biochemists, the latter were smugly ignorant of the powerful forces the devotees of the new approach were mobilizing to change the way biological problems would be thought about and tackled in the coming decades. In this connection, I should note that the elucidation of the initial energy input step in protein synthesis came not much more than a year after Watson and Crick's revelation of the structure of DNA. I recall with some embarrassment my perfunctory first reading of their momentous paper in 1953. It is true that the paper appeared as I was returning to the Huntington from Lipmann's lab, preoccupied with the mechanics of protein synthesis and immersed in the juices of cells and in biochemical energetics. But my neglect of its implications was rooted in the kind of parochialism that affected both camps.

The two streams of inquiry were converging faster than any of us realized, however. The molecular biologists could not postpone much longer their confrontation with the black box of cell machinery, and the biochemists were pushing up against the implications of information processing and big-molecule structure. Ribosomes, rich in RNA, were somehow converting sequence information from DNA into amino acid sequences in proteins. How? Our next discovery at the Huntington fit a very large piece into this translational puzzle—transfer RNA. It completed the bridge from biochemistry to molecular biology.

After discovering amino acid activation, I had jubilantly dashed upstairs and told Fritz Lipmann of my findings. Before I knew what hit me, he put a young associate to work on the obvious next step of isolating and purifying one of the many amino acid–activating enzymes presumed to be in the soluble fraction of the cell. (It seemed probable that each of the amino acids would have to be activated by a separate enzyme.) Lipmann sent another associate to the Huntington lab to glean details of our techniques for studying protein synthesis and set him to work on the problem without informing us.

While I appreciated the openness and competition inherent in science, I resented having the baton unceremoniously ripped from my sweating hand before I had even finished the first lap. Part of this feeling was, I now believe, based on what I saw as my amateurism, or unprofessionalism: a suspicion that I was not prepared to compete in the performance of the next steps, the building of the necessary ramparts of proof to substantiate fully such a major discovery. I knew that one could not expect to be a serious scientist without accepting the reality that intense competitiveness and the giving and receiving of merciless criticism were essential to the enterprise, but it nevertheless got me down. I began to find scientific meetings tiresome; the first day or so I would be euphoric, and then I'd get depressed. I have since learned this is not an uncommon experience among scientists, but in those days I often felt alone. It seemed to me that many of my colleagues were positively ecstatic at the prospect of beating other scientists to the draw, often by doing experiments they knew their competitors were planning to do. I was easily hurt, too, by instances in which my colleagues failed to acknowledge what I saw as a clear indebtedness to our group's work. I felt the urge to skulk away and find a problem that nobody else would want to work on. The trouble was that what was hot was also competitive.

Still, there was much to do. I continued to pursue aspects of the activation reaction. We were all excited by the wealth of knowledge about protein synthesis that had been unearthed at the Huntington in a relatively short time. We had become more proficient in the art of gently breaking cells open so that their ribosomes continued to make respectable amounts of protein. And now we had pinpointed one important thing the soluble fraction was doing:

its indigenous enzymes were activating amino acids by using energy from ATP in the linkage between Ad · P and P ~ P.

The big question, to which we had no inkling of an answer, was, How can an amino acid chain–making system sort out which amino acids are needed for a particular protein and in what order they are to be put? Our vague thinking was that the selection and ordering of amino acids was done by complex processes on the ribosomes, which would emerge in due time, as we learned more about their structure and function. Ribosomes are made up of proteins and RNA. RNA is chemically very much like DNA: it consists of four nucleotides—adenylic acid (Ad · P), guanylic acid (Gu · P), cytidylic acid (Cy · P) and uridylic acid (Ur · P, very similar to thymidylic acid in DNA)—linked together by their sugar-phosphates bonds. The sugar in RNA (ribose) is slightly different from that in DNA (deoxyribose), giving RNA somewhat different chemical properties.

John Littlefield, ca. 1980 *(courtesy John Littlefield)*

We assumed that at least some of the ribosomal RNA in a cell was a copy of the DNA. That must be so if DNA, as the only source of heredity information, sequestered in the nucleus, was to guide the manufacture of protein on ribosomes out in the cytoplasm. The sequence of nucleotides in ribosomal RNA, we thought, must somehow provide a template that would select and order the amino acids to be put into protein molecules as they were being made.

When John Littlefield joined the Huntington group in 1954, Paul suggested he look into the possibility that the system that made protein in cells might also make RNA. Paul bet it would and taped a nickel on the door of John's lab to back up his hunch. Not much was known about how RNA was synthesized. Paul speculated that if the amino acid–activating enzymes were making Ad · P ~ amino acids, these activated molecules might in a sense be double-barreled—that is, able to donate their "energized" amino acid to a protein-polymerizing machine *or*, alternatively, able to donate their Ad · P portion to an RNA-polymerizing machine. If other enzymes could similarly link Gu · P, Cy · P, and Ur · P, RNA (Ad · P ~ Ur · P ~ Gu · P ~ Ad · P ~ Cy · P, and so on) would be made. It was a long shot.

John got involved in another project while the nickel stayed neglected on his door. Paul and his associate Mary Louise ("Sis") Stephenson finally decided to test the idea themselves a year later, in late 1955.

Since the Ad · P portion of ATP is one of the four key constituents of RNA, Paul and Sis incubated radioactive ATP (its Ad part labeled with ^{14}C) with ribosomes and soluble fraction to see if any radioactive Ad was incorporated into RNA. The result was encouraging. The RNA isolated from the system after incubation was indeed radioactive. They were pleased but suspicious: Could the radioactive Ad · P, or even the whole Ad · P ~ P ~ P molecule, just be sticking onto the RNA somehow, and not really built into it? How about a control experiment using some other radioactive compound that was presumed *not* to get built into RNA? Having plenty of radioactive amino acids available, they decided to try one of them. The experiment with radioactive Ad · P ~ P ~ P was repeated, this time accompanied by a parallel incubation of an identical system with radioactive amino acid instead. The result was that radioactivity from ATP was again incorporated into RNA but that, in the parallel incubation, amino acid also became firmly attached to the RNA.

Most of the RNA in these experimental systems was contained in ribosomes. But there was a small amount of RNA in the soluble fraction as well. Paul and Sis therefore repeated the experiment once again, this time adding test tubes in which ribosomes alone and soluble fraction alone were exposed to the radioactive compounds. The RNA of the soluble fraction turned out to be much more highly labeled than that of the ribosomes. Surprise compounded!

The small amount of RNA residing in the soluble fraction, about 10 percent of the total RNA of the cell, had been noticed before, but nobody understood its significance. It was assumed to be "junk"—perhaps fragments of the larger RNA of ribosomes left over from the process of breaking open the cells. Indeed, soluble-fraction RNA was much smaller in molecular weight than ribosomal RNA (and so remained soluble when the ribosomes were centrifuged down). Here, however, was unexpected evidence that this soluble RNA might have some very special function.

At this point, in January 1956, Paul stopped doing experiments. He had just been appointed to succeed Joseph Aub as head of the Huntington laboratories and professor of oncologic medicine at Harvard. New duties pressed. He had not given up his clinical duties either, which also made demands on his time. And a postdoctoral associate, Lisa Hecht, who was experienced in the technology of RNA and who he had hoped would pursue the work, had delayed her arrival by several months. It was not until June that Paul told me of the five-month-old findings, in his usual low-key, half-puzzled, half-deprecating way. I was enormously intrigued by what was clearly a direct interaction between amino acids and RNA—between the building units of protein and what we presumed to be genetic information–bearing molecules in the very fraction that I had found carried out the first step of protein synthesis.[2] Paul was happy to have me pursue the lead.

I repeated and confirmed the experiments Paul and Sis had done. Since the fraction active in attaching amino acids to RNA was the same as the one that activated amino acids, we postulated that ATP and amino acids were reacting to form the activated intermediate (reaction 1, chapter 4) and that then the activated amino acid was being passed on to the RNA thus:

(ATP)

Ad·P~P~P

(adenyl-amino acid)

$$\begin{matrix} & & Ad{\cdot}P{\sim}CO & & RNA{\sim}CO \\ COOH & & | & & | \\ | & \rightleftarrows & R{-}CH & \rightleftarrows & R{-}CH \\ R{-}CH & & | & & | \\ | & & NH_2 + RNA & & NH_2 \\ NH_2 & & & & \end{matrix}$$

(amino acid)	P~P (pyrophosphate)	Ad·P (adenylic acid)

For each of the amino acids, these events were presumed to be occurring on the surface of a specific enzyme.

Further experiments quickly showed that ATP was specifically required to make the new compound (the substitution of CTP, UTP, and GTP would not do) and that the amino acid was linked to the RNA in a firm chemical bond. Significantly, too, we found that incubation of the new compound (RNA-amino acid) with the soluble fraction containing the enzymes, and extra P ~ P, caused the radioactive amino acid to disengage rapidly from the RNA. This suggested that P ~ P pushed the equilibrium backward—to the left—to form ATP and free amino acid.

The exciting implication of the phenomenon that stared us in the face was that we had found in this small amino acid–binding RNA a key intermediate in protein synthesis—a molecule or molecules that accepted activated amino acids and carried them to the ribosomes. (Another possibility—that soluble, cytoplasmic RNA acted as a kind of a waste disposal unit, picking up unwanted amino acids—was deemed unlikely and rejected.) That such a hypothetical intermediate might have a role in the mysterious process of ordering amino acids, we had no doubt, and no evidence. A crucial first test was obvious: incubate the soluble fraction containing the activating enzymes and indigenous soluble RNA with ATP and radioactive amino acids, reisolate the fraction with amino acids now attached to RNA, and reincubate it with ribosomes to see whether the radioactive amino acids left the RNA and appeared in protein molecules.

Trying hard to suppress my excited sense that this experiment was going to raise the curtain on a magnificent new vista in the drama of biology, I put the ingredients together with more than the usual tender care. I checked off each micromeasured component as it entered the tubes: the soluble fraction (containing activating enzymes and soluble RNA), ATP and radioactive amino acid, plus certain other necessary goodies. I put the mixtures into a body-temperature water bath and let them incubate exactly ten minutes, then stopped the reactions by putting the tubes into ice water. (Enzyme-mediated reactions are temperature dependent: cold essentially stops them.) To remove the ATP and radioactive amino acid, I made the proteins and RNA insoluble with a mild acid, centrifuged them down, and discarded the fluid containing ATP and amino acid. I saved a sample of this RNA-protein fraction to measure the amount of radioactive amino acid on the RNA. Lastly I set up a series of new incubations in which the reisolated soluble fraction, brought back to neutral pH, was incubated with ribosomes and GTP, the nucleotide Paul and Betty Keller had found to be necessary in later stages of protein synthesis (see page 74). Some identical sets of ingredients I left in the cold, unincubated, and from some I omitted the GTP. (I omitted ATP from all the tubes of the second incubation on the assumption that it was no longer needed, because the amino acids were already activated.) I stopped all the reactions with strong acid, separated the protein and RNA from each tube, and spread them all out on plates to dry prior to measuring them with a Geiger counter.

It was night by the time the samples were dried, stacked, and ready to move automatically under the counter tube. I still can clearly see the dark windows of the lab, smell the organic solvents, hear the buzzing of a defective fluorescent lamp in the next room. In front of me were the transfixing flashing lights on the Geiger counter as the samples began to be counted. First, the unincubated complete system: 489 counts of labeled amino acid in RNA, 30 counts in protein. Good—the few counts in protein are expected background in such a relatively crude system. Next, the incubated system, GTP omitted: 111 counts in RNA, 40 counts in protein. Okay—RNA loses amino acid, but it is not getting into protein. Next, the complete incubated system: 180 counts in RNA, 374 counts in protein. Those little numbers caused a shiver to go down my spine: amino acids had left the RNA and entered protein! From that night on,

I had little doubt that this small soluble RNA was the physical link between activated amino acids and their ordered arrangement in protein molecules.

Sis, Paul, and I spent the remaining months of 1956 and the early ones of 1957 in pinning down the evidence that "soluble RNA" was indeed a key intermediate in protein synthesis. ("Soluble RNA" was soon to be renamed *transfer* RNA [tRNA], as its function became better established.) We confirmed that the transfer of amino acids to RNA was dependent on GTP and showed that it was a relatively rapid process compared with the rate of incorporation of amino acids into protein. We separated transfer RNA to which we had attached amino acids from the soluble fraction, added fresh soluble fraction and GTP, and showed that the amino acids were again almost completely transferred to protein. By various chemical and physical treatments, we satisfied

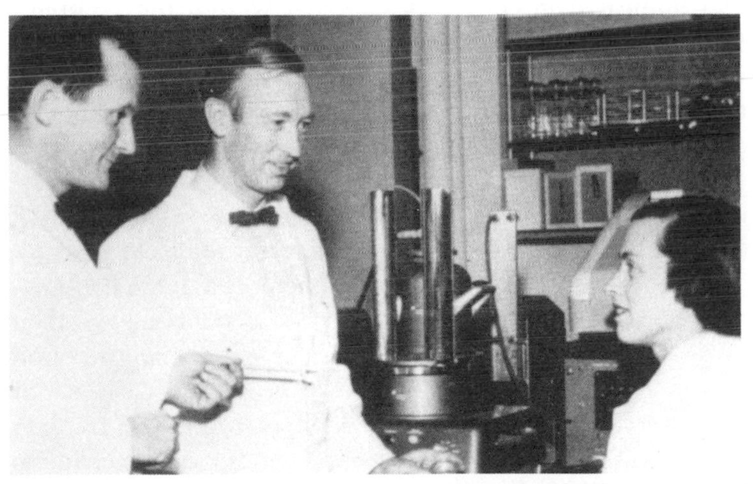

Mary Louise Stephenson *(right)*, Paul Zamecnik *(middle)*, and author, 1955 *(Massachusetts General Hospital archives)*

ourselves that the amino acids were chemically bonded to the RNA. Moreover, the amino acids reacted readily with that useful agent for trapping activated amino acids, hydroxylamine. This indicated that the energy level in the new association between RNA and amino acids was approximately equivalent to that between Ad · P ~ and amino acids in the compound from which it appeared to be formed.

In addition, I found that as I added more and more *different* amino acids to the system, there was room for all of them. This had to mean that each of the twenty amino acids was separately bound to some site on the RNA molecule(s). Transfer RNA must, then, be either a single species of molecule with binding sites for all twenty of the amino acids or twenty (or more) different RNAs, each with a specific binding site for each amino acid. The latter proved to be true, but we could not make the distinction then. All of us at the Huntington were convinced we had uncovered the second major step in the cell's machinery for making protein.

I wrote up a preliminary account of these findings, which was published in early 1957.[3] Again we felt our competitors' hot breath on the backs of our necks. Colleagues in Sweden were also picking up evidence suggesting that amino acids passed into a state different from that of free amino acid before they entered protein. Kikuo Ogata at Niigata University in Japan, Paul Berg at Washington University, and Robert Holley at Cornell University were also finding signs of steps in which amino acids were bound to RNA.[4]

What justified our inference that transfer RNA had anything to do with determining *order?* The evidence was then only circumstantial: if DNA carried genetic information specifying amino acid sequence, and if RNA carried out DNA's role in the cytoplasm, which everyone was by then assuming it did, then the direct chemical association of amino acids with RNA *must* have implications for sequencing.

At the end of 1956, I had my first visit from a card-carrying molecular biologist. Jim Watson had just become a professor of biology at Harvard and was probing the structure of the ribosome. He had heard rumors of our discovery of transfer RNA, and I eagerly told him of our findings. He was restlessly attentive and, when I had finished, told me that Francis Crick had forecast the existence of transfer RNA–like molecules a year or so earlier. He wondered whether I had heard of the adaptor hypothesis. I was astonished and a bit miffed that I had not. Jim explained that Francis had been theoretically wrestling with the question: if the machinery of protein synthesis contains information specifying the order of amino acids in protein—presumably, in the form of DNA-derived RNA acting as the ordering template—how would the template "know," or

James Watson, ca. 1985 *(Cold Spring Harbor Laboratories archives)*

chemically recognize, an amino acid if it bumped into it? This concern was a natural extension into chemistry, into molecular structure, of Francis's fascination with the genetic code. There just was no trace of chemical homology or complementarity between amino acids on the one hand and RNA on the other. Suppose, thought Francis, an enzyme first attached each amino acid to a special, unique short length of RNA. The amino acid would then have had conferred upon it a chemical identity that *could* be recognized by an ordering RNA template. How could this recognition be accomplished? The same way as in the double-stranded DNA molecule, where one strand recognized the other by forming hydrogen bonds between adjacent pairs of nucleotide bases whose shapes are complementary to one another. Francis Crick envisioned that these postulated small molecules, which he called adaptors because they adapted amino acids so that they could be recognized by an ordering template, would be really quite small, perhaps only three to ten nucleotides in length. It was postulated that three nucleotides would be the minimum number of bases needed to code for a single amino acid. (The theory glossed over the problem of how the system would "know" how to attach the amino acid to the correct RNA adaptor; it vaguely relegated the job to the versatility of an enzyme.)

I was bowled over by the ingenuity and beauty of Francis's idea and sensed that it had to be the explanation of our experimental findings. An image arose in my mind: we biochemical explorers were hacking our way through a dense jungle to discover a beautiful long-lost temple, while Francis Crick, flying gracefully overhead on gossamer wings of theory, waited for us to see the goal he already was gazing down upon.

Francis apparently conceived the adaptor hypothesis late in 1954. In early 1955, he wrote up the idea and sent it out to friends: "On Degenerate Templates and the Adaptor

Hypothesis: A Note for the RNA Tie Club."* It was never formally published. This remarkable paper begins by quoting a little-known medieval Persian writer: "Is there anyone so utterly lost as he that seeks a way where there is no way?" It goes on,

> I don't think that anybody looking at DNA or RNA would think of them as templates for amino acids were it not for other, indirect evidence.
>
> What the DNA structure *does* show (and probably RNA will do the same) is a specific pattern of *hydrogen bonds,* and very little else. It seems to me, therefore, that we should widen our thinking to embrace this obvious fact. Each amino acid would combine chemically, at a special enzyme, with a small molecule which, having a specific hydrogen-bonding surface, would combine specifically with the [RNA] template. This combination would also supply the energy necessary for polymerization. In its simplest form there would be twenty different kinds of adaptor molecule, one for each amino acid, and 20 different enzymes to join the amino acid to their adaptors. Sydney Brenner, with whom I have discussed this idea, calls this the "adaptor hypothesis," since each amino acid is fitted with an adaptor to go on to the template.
>
> The usual argument presented against this latter scheme is that no such small molecules have been found, but this objection cannot stand.

* The RNA Tie Club was founded in Berkeley, California, by the physicist turned theoretical biologist George Gamow, in about 1953. Its aims were to "solve the riddle of RNA structure and to understand the way it builds proteins." The club's emblem was an RNA tie "as designed by George Gamow, to be produced by an appropriate haberdasher in Oxford, England." There were to be twenty members, one for each amino acid, and four honorary members, one for each base in RNA. The idea was that key experimenters and theoreticians would write and discuss in a congenial atmosphere speculative biological papers not yet ready for publication.

By early 1956, Francis had been persuaded by his friends that he should get the adaptor hypothesis into print. He spoke at a London meeting of the British Biochemical Society in February 1956 and in the published proceedings again related the theory, adding the suggestion that the adaptor might be a small length of RNA.

In the summer of 1956, Francis gave an informal talk on ribosome structure and the adaptor idea at a Gordon Conference in New Hampshire, not far from us in Boston, in the evening at the end of a five-day meeting. Paul Zamecnik had attended that meeting but had left early and had not heard Francis's paper. How could we have missed learning of the idea before Jim told us? Looking back, I think the explanation lies in those flaws in communication between specialists of dissimilar orientation that I alluded to earlier.[5]

At any rate, I am constrained to add that Crick's adaptor idea, while one of the more brilliant insights in molecular biology, stands more as a monument to man's imagination than as the springboard to a major discovery. This is, of course, what we generally expect of a good hypothesis. In this case, we biochemists could assert with pride that by dint of smaller imaginative jumps, hard labor, and considerable luck we had unearthed tRNA and appreciated its significance for determining order without the aid of the hypothesis—even if its promulgation did percede our discovery. In this instance, a grand theory neither substituted for nor guided the successful analytical dissection of the machinery of protein synthesis.

The meeting with Jim Watson symbolized the beginning of the end of the barrier between traditional biochemists and molecular biologists. I believe that the discoveries of amino acid activation and particularly of transfer RNA played a large role in that felicitous turn of events. From then on, we all knew we needed one another.

I busied myself in early 1957 with further experiments to refine our basic findings in collaboration with Jesse Scott and Lisa Hecht, both Huntington colleagues expert in RNA biochemistry. We completed a definitive publication that belatedly emerged in the *Journal of Biological Chemistry* in March 1958.[6]

I also completed a pretty little experiment that Paul and I had discussed with Paul Boyer at the University of Minnesota. The postulated mechanism of amino acid activation involved the linking of the carboxyl group of amino acids to the innermost phosphate of ATP (with the ejection of P ~ P), as we have seen. In the presence of the amino acid acceptor hydroxylamine, the Ad · P ~ amino acids so made will react to form the hydroxamic acids of each of the amino acids, releasing Ad · P. This reaction is chemically equivalent to reacting the Ad · P ~ amino acid with another amino acid to form a dipeptide. If the mechanism was in fact as we had postulated, the carboxyl *oxygen* of the amino acid should be found in the released Ad · P (see page 78, where the critical oxygen is depicted in bold type). Boyer knew how to make amino acids whose carboxyl group contained a heavy isotope of oxygen (^{18}O). I asked Lipmann, upstairs, to supply a bit of the activating enzyme for the amino acid tryptophane, which his group had purified. I ran the reaction with ATP, hydroxylamine, and the ^{18}O tryptophane and sent Boyer the resulting Ad · P. To our delight, the oxygen had indeed been transferred to the Ad · P, thus giving strong support to our mechanism for the activation reaction.[7]

In the spring of 1957, Francis Crick came to visit us. He was elated. The discovery of an RNA species that apparently performed the task of his postulated adaptors prompted him to enter the experimental fray. He invited me to spend a year in Cambridge to help him set up a lab and try his hand at experimental work. I gladly accepted his invitation.

Before returning to the Huntington from Denmark in 1953, I had applied to the Americana Cancer Society for a prized five-year salary called a scholar award. These awards are granted to candidates who emerge victorious from a rather odd competitive process. The applicants—I do not know how many there were initially—were weeded down to twelve by way of a review of submitted credentials. These twelve were invited to spend a weekend at Arden House—an elegant French Provincial mansion bequeathed to Columbia University by the Harrimans. The weekend was spent with a committee of amiable scientists and academic physicians who, directly and indirectly through a series of outdoor walks, elegant meals, and general bull sessions, plumbed the depths of our personalities and intellects, a process that weeded out half of us. I was one of the happy six who were finally chosen.

The last year of the scholar award coincided with the proposed academic year in Cambridge. I had been assured of a faculty post at the Huntington upon my return. The pieces of the puzzle of protein synthesis were falling into place, but two big ones were still missing. How was the language of the gene inscribed in the sequence of bases in DNA so as to specify the sequence of amino acids in protein? And how did that information get to the ribosome and program it for translation? Both pieces would be found in the next three years.

6

❖ ❖ ❖

Messenger RNA

Cambridge, England, and more particularly a few small offices and laboratories in some temporary buildings, erected during World War II, in the quadrangle of the venerable Cavendish Laboratory, was in 1957 the world's informal communications center for the young science of molecular biology. The concentration of scientific talent at the Medical Research Council Unit for the Study of Molecular Structure of Biological Systems, as the complex was known, was awesome. The unit's head, Max Perutz, and his associate, John Kendrew, would win the Nobel Prize in chemistry in 1962 for their revelations of the complex structure of proteins. That same year's Nobel Prize in physiology (medicine) would go to Francis Crick and James Watson for their solution of the structure of DNA in 1953. There also was Sydney Brenner, a South African virologist, second only to Francis in glitter of mind (and now director of the MRC Unit), and Vernon Ingram, who was then developing the technology to "fingerprint" proteins that was to become valuable in characterizing the molecular basis of genetic disease. The microbial geneticist Seymour Benzer, developer of new precision techniques for dissecting genes, was visiting for a year, as was another American geneticist,

George Streisinger. Paul Doty of Harvard and Alexander Rich of MIT, noted investigators of molecular structure, were there for several weeks during my time there. And Frederick Sanger, a pioneer in the emerging technology for determining the sequence of amino acids in proteins, was an MRC neighbor and frequent visitor. He was awarded the Nobel Prize for his work that very year. Within the next few years, several of these men would play key roles in further elucidating the three-dimensional structure of proteins, in proving the parallel (colinear) relation between bases in DNA and amino acids in protein, in defining the molecular basis of inherited disease, and, as I shall relate, in the discovery of messenger RNA.

Francis Crick was the pied piper and poet of biology and its dominant figure throughout the 1950s. Few who knew him could fail to be impressed by his grace and eloquence, the breadth and discrimination of his knowledge, his wide-ranging and disciplined imagination. Conversations led by him would continue much of the day—at morning tea, at

Francis Crick, ca. 1975 *(courtesy Francis Crick)*

lunch, at afternoon tea, at nearby pubs in the early evenings, and often at dinner—and touch on cellular information processing, coding, molecular structure, colinearity, protein synthesis, theater, books, politics, and personalities. Francis stated and clarified propositions, sifted data, rejected trivia. He was sharp in criticism but charitable and amusing. He had an uncanny ability to find his way through a welter of confused information to the nub of the issue. Sydney Brenner was a sort of Crick lieutenant, adding force and spice and his own insight to the argument. Together, they had a remarkable ability to keep a subject clearly defined, on track, and moving forward. Overall, an atmosphere of exhilaration, expectation, and humor pervaded the MRC.

Francis dealt with a prodigious volume of correspondence from scientists all over the world who sought his views, interpretations, suggestions for further experiments, or just plain approval. He replied thoughtfully and constructively to most of those letters, often after a lively dissection and discussion of the more provocative ones in our lab powwows.

Francis had switched from the war-related study of physics in London—in which he had not quite obtained his doctorate—to biology at Cambridge in 1947, at the age of thirty-one. Like others who had turned to biology in the 1940s and 1950s, Francis had been influenced by the remarkable little book *What Is Life,* by Erwin Schrödinger, a founder of quantum mechanics. He wrote in a fellowship application to the Medical Research Council in 1947,

> The particular field which excites my interest is the division between the living and the nonliving, as typified by, say, proteins, viruses, bacteria and the structure of chromosomes. The eventual goal, which is somewhat remote, is the description of these activities in terms of their structure, i.e. the spatial distribution

> of their constituent atoms, in so far as this may prove possible. This might be called the chemical physics of biology.[1]

Then and there, he defined with astonishing prescience molecular biology, a term that would not emerge for another decade. He was still working for his degree in 1953 when he and Watson—who was only twenty-four and *had* his Ph.D.—discovered the structure of DNA. Francis's engaging account of his life in science illuminates the origins of molecular biology and his role in its development.[2]

An important facet of the grand quest during the 1950s for an understanding of protein synthesis was the coding problem: which sequences of nucleotides in DNA (and thence in RNA, assuming it to be a copy of DNA) represented, coded for, each of the amino acids in protein? That genetic information must reside in the *order* of the nucleotides was implicit in the four-unit linear polymer structure of DNA. There was reason to suspect that the minimum *number* of nucleotides in DNA needed to code for each of the twenty amino acids was *three:* one per amino acid was obviously not enough; two was inadequate, as there were only sixteen possible combinations of the four nucleotides taken two at a time; a group of three, however, gave more than enough unique combinations of sequences to code for twenty amino acids, namely, sixty-four.

The physicist George Gamow, inspired by Francis and Jim's revelation of the structure of DNA, had a provocative idea for a code. He speculated that short sequences of paired nucleotides along the double helix of DNA might create real physical cavities or indentations along its length that were sufficiently different from one another to accommodate each of the twenty amino acids, thereby ordering them.

However, the lack of any chemical resemblance or complementarity between DNA bases and amino acids made Gamow's idea untenable. Still, it had the merit of pushing people, notably Francis and Sydney Brenner, into pondering coding. This led them to some ingenious ideas, and eventually to the adaptor hypothesis. That idea, as we have seen, had the effect of switching the problem of "recognition" of nucleotide sequence by the amino acid–ordering system from nucleotide–amino acid interaction, à la Gamow, to nucleotide-nucleotide interaction (between tRNA and postulated template RNA). This made sense chemically and had the added merit of turning out to be right.

Aside from his having foreseen it, Francis had another reason to be excited about our discovery of transfer RNA. It seemed obvious that if the protein-making machinery of the cell actually linked individual amino acids to distinct species of RNA molecules, we had a direct chemical handle on the cracking of the genetic code. Was it possible to isolate individual tRNA species and identify the nucleotide sequence responsible for binding a particular amino acid? There were two very big stumbling blocks, however. First, the tRNA fraction we had isolated was a mixture of many specific tRNAs, and the formidable task of separating it into individual species would take many more years to accomplish. Second, tRNAs were unconscionably and uncooperatively large molecules. Instead of being a few nucleotides in length, as Francis had imagined in the adaptor hypothesis, they were some eighty nucleotides long. Searching for three coding nucleotides among eighty was a daunting project. For some time, Francis even doubted that our tRNA molecules *could* be his adaptors. He wondered, lamely, if perhaps "real" adaptors might be derived from, or snipped off of, tRNAs in the cell once amino acids had become attached to them. Nature's needs, however, were more sophisticated than Francis's ideas; the whole complex tRNA molecule was essential, both to identify its enzyme-bound activated amino acid for purposes of linking the amino acid to it and to interact with the ribosome to ensure proper positioning of the tRNA on a then unknown but logically required template.

But Francis still wanted to get his hands dirty, to handle some tRNA and to taste experimentation, beginning modestly with isolating tRNA from rat liver and making some preliminary efforts to separate it into different amino acid–binding species. We set up a lab in the nearby Molteno Institute with the help of my friend and colleague John Littlefield, who was then visiting in Cambridge, too, and we started to work. We would do an experiment and get some variation in results that Francis felt obliged to analyze and ponder at length. I would assure him that the variations were very likely an error—we would not find them if we repeated the experiment. Or, we would try some new technique for separating different tRNAs, and things would go wrong. I was used to all that, but it soon became apparent that the great theorist was getting bogged down in triva. I felt he was experiencing a wistful nostalgia for juggling big ideas, not test tubes. Watching Francis in this context was fascinating. He had an uncanny ability to analyze and criticize, *in detail,* the experiments of others, but at the bench he became mired in the day-to-day messiness and inconclusiveness. The raw face of biochemical experimentation was not right for his mountaineering spirit. The morning I arrived to find him on his knees on the floor trying to find a white rat that had turned gray with dust while eluding efforts to catch it was the moment we both acknowledged that his experimental days were over.

Another of the big intellectual-technological challenges that year, and for a few more years to come, was the proof of the sequence hypothesis—the demonstration that sequential, single-base alterations of the nucleotides in DNA (in other words, mutations)—caused *colinear* alterations of amino acids in protein molecules. Put another way, if a gene is a stretch of DNA specifying an amino acid chain, does the location of a change in a nucleotide base in the DNA chain coincide linearly with a change in an amino acid in the protein? Since it was not then possible to isolate and make multiple copies of—to clone—individual genes in order to analyze their base sequences by chemical means, genetic techniques had to be used. Before the advent of fine-structure genetics, developed by Seymour Benzer in the mid-1950s, genes had been "mapped"—that is, their relative positions along the chromosome had been determined—by recombination, using the techniques developed by the pioneering geneticist Thomas Hunt Morgan. This involved mating organisms with different genetic traits, either naturally occurring or induced by mutagenic agents, and measuring the frequency with which the traits recombined in the progeny. The more frequently two traits reappeared, the greater the chance of recombination between them and so the farther apart the genes must be on the chromosome. In this work, it was assumed that genes themselves were somehow inviolate and indivisible, that recombination occurred only *between* genes, not within them. This assumption was erroneous simply because methods and experimental subjects, such as Morgan's fruit flies, did not permit the detection of extremely rare recombinational events between short segments within genes. By using viruses that infect bacteria (bacteriophages), Benzer developed an ingenious technique, called complementation, for defining a gene. Then, by "crossing" mutant viruses that had single-base changes in their DNA, he was able to detect many extremely rare recombinational events *within* single genes.

Benzer's genes were defined by a functional test—their protein product had not yet been discovered. To get at colinearity, it was essential to have both a gene and the protein product of that gene in order to see whether mutations in DNA did, in fact, coincide linearly with amino acid alterations in protein. Fred Sanger had begun to work out methods for determining amino acid sequence in protein. So, the time was ripe for an ingenious investigator to pick the right system.

The winner of the race to prove colinearity was Charles Yanofsky at Stanford University, with Sydney Brenner close on his heels. Yanofsky chose as his protein an enzyme in the bacterium *Escherichia coli*. With prodigious effort, he identified and located most of the amino acids in the protein's chain, which is about three hundred amino acids long. By crossing *E. coli* organisms that had different mutations in the gene for the enzyme, he "mapped," à la Benzer, many single-nucleotide changes in the gene. The gene was estimated to be some one thousand nucleotides in length. The positions of mutations in the DNA and of consequent amino acid changes in the protein turned out, in fact, to be colinear. Yanofsky's work was so precise that he could further affirm that each amino acid was, as had been guessed, coded for by approximately three nucleotides (one thousand nucleotides in the gene, three hundred amino acids in the protein).

For me, the colinearity issue that year was not as exciting as the remaining mysteries of the protein synthesis mechanism. The work on colinearity completely bypassed the black box of the machinery. While matching DNA bases with amino acids involved high technical ingenuity—not until the early 1960s were the proofs of Yanofsky and Brenner final—it was a matter of nailing down the obvious.

Contrast this with the situation regarding protein synthesis. We had genetic information in DNA; we had the ribosome factory with lots of RNA in it; we had transfer

RNA molecules conferring upon amino acids an RNA-recognizing capacity; and we knew how the whole process was energized. But a tantalizing piece was missing: how did the DNA information get to and program the ribosome? DNA was generally (except in bacteria) sequestered in the nucleus of the cell, and the ribosomes functioned outside in the cytoplasm. There had to be some direct *physical* communication between them.

The problem had bothered Paul Zamecnik almost as soon as Francis and Jim's model of DNA was launched. He wrote,

> In the summer of 1954, we inquired of our scientific neighbor, Paul Doty, as to how the RNA of the [ribosome] could serve to order the activated amino acids. Dr. Doty mentioned that a Dr. James Watson was visiting him, who with a colleague, Dr. Crick, had recently proposed a model structure for DNA. He would send Dr. Watson over to visit me and I could ask him that question. I looked at Dr. Watson's young face above his white Irish turtle neck sweater, then at his wire model of the double-stranded DNA, and inquired how the message of the DNA made its way into the sequence of protein. Unfortunately, it seemed to me privately, the bases were facing in, rather than out. . . . Was protein made directly on a DNA template? Probably not, it appeared, because there was no DNA in the [ribosome]. Was RNA made on the DNA template? No answer to this question either, although it seemed likely. How did this complicated double helix unwind? There was a gulf between DNA and protein synthesis, Dr. Watson agreed with a diffident smile, as we parted and he took off on vacation to look at birds.[3]

Those of us who worked with slow- or nongrowing animal cells were comfortable envisioning the ribosome as a stable factory already containing an RNA transcript of DNA.

These cells were relatively stable protein builders, their ribosomes were loaded with RNA, and we assumed that base sequence information in ribosomal RNA would eventually be shown to reflect the information in DNA—in other words, that the ribosomal RNA was the likely template. Rat liver and tumor cells did not yield any insight into one critical parameter of protein synthesis: namely, what happened when it was called upon to *change* its rate rapidly. It was a study of shifts in protein synthesis, induced in a simpler, rapidly growing organism, that disclosed the last great mystery of the mechanism of protein synthesis.

In January 1958, I stopped off in Paris to visit colleagues at the Pasteur Institute at a critical moment. Jacques Monod and François Jacob were in the early phases of the brilliant train of experiments that led by the end of the decade to profound new insights into how bacteria control the expression of their genes—how they turn on and off a gene's production of protein. Arthur Pardee, an accomplished American microbiologist and biochemist, had joined the Pasteur group in the fall of 1957 and by the time I arrived had already obtained the early results of an experiment that would make everyone in protein synthesis uncomfortable for the next two years—until the light came on. The experiment has come to be known as PaJaMo (after Pardee, Jacob, and Monod).

At Jacques Monod's suggestion, Arthur began by investigating the expression of the gene that prescribes the amino acid sequence of the enzyme β-galactosidase, an enzyme that helps bacteria use, or break down, the sugar lactose. Monod had delineated the β-galactosidase system, and François Jacob had done much to reveal the nature of bacterial mating or conjugation. The PaJaMo experiment combined the skills and imaginations of all three men. When certain types of bacteria are mixed together, they simultaneously mate, or conjugate, and the DNA of the donor type enters the recipient type at a steady and fixed rate. Thus, the entry of any particular gene can be timed precisely in a large population. If the donors have an intact gene for β-galactosidase—meaning, they can make the enzyme protein normally—while the recipients are genetically incapable of making the enzyme, the experimenter can ask, How fast can the gene program the synthesis of the enzyme expressed after it enters a recipient cytoplasm that had no prior "knowledge" about how to make the enzyme? How long will it take a bacterium's uneducated protein synthesis machinery to reach the maximum rate of protein production upon being presented with the necessary blueprints? What the recipient cells were doing at any chosen moment after the start of mating could be determined simply by interrupting the mating process—by separating the mating pairs by dumping the whole population into a blender—and then measuring the amount of enzyme present. (An ingenious method was used to be sure that any enzyme measured after mating was enzyme made by recipient bacteria, not by donors: the experimenters used donor cells that were streptomycin sensitive and recipient cells that were streptomycin resistant; after mating, the antibiotic was added to kill off all the donors.)

The surprising, and for a long time confounding, result of this experiment was that mated recipient bacteria began making β-galactosidase at the maximum rate *without any significant delay* after the gene entered their cytoplasm.[4]

It was known from the work of many laboratories that bacteria contained essentially the same machinery for making protein as did animal cells: activating enzymes, ATP, GTP, transfer RNAs, and ribosomes. The PaJaMo experiment indicated that genetic information immediately triggered protein synthesis in a system apparently already poised to do the job. Since the ribosome factories were big, complex structures composed of a lot of RNA and protein, it was almost inconceivable that *new* ones could have been made quickly enough to account for the rapid onset of protein synthesis seen in the PaJaMo experiment.

Other things were becoming known about ribosomes as well. For instance, most of the RNA they contained did not look like information for protein synthesis. If ribosomes were making all the protein in a cell, and if ribosomal RNA was a template upon which these many proteins were being made, one might have expected a large variety of RNAs, but that was not found. Ribosomal RNAs came in two big chunks, both much bigger than needed for a template for the average protein. Second, one might expect ribosomal RNA to bear some resemblance, in terms of its nucleotide composition, to the DNA of a cell. It did not. These findings, together with the implication of the PaJaMo experiment that a cell could make new protein without making new ribosomes, left all investigators in the field baffled. In those two winter weeks in Paris, Pardee and I took long walks, exploring that beautiful city and ruminating on protein synthesis. Jacques Monod humorously needled me about the scheme of protein synthesis with which I was identified. He could not help wondering if bacteria, unlike animal cells, might somehow be making protein directly from DNA. Our present understanding of the universality of biochemical mechanism was not then as firmly established as it became a few years later.

Jacques Monod, incidentally, could have talked me into anything. Like Linderstrøm-Lang, he had the grand dimensions of a hero of our time. He frequently risked his life in the French resistance during World War II. He was a musician, mountain climber, sailor, and philosopher. He was gracious and provocative. An idealist, with a prodigious breadth of vision, a dramatic flair, and a monumental ego, he was, after Francis Crick, biology's most fertile theorist. In addition, unlike Francis, he was a brilliant experimentalist. I got to like him immensely, and we shared an enthusiasm for our work, a love of cruising under sail, and a romantic way of looking at the world. I admired him, too, as I did Francis Crick and François Jacob, for his open, eloquently articulated rejection of maudlin mysticism and vitalism—for being explicit about his atheism. All three men were awed by the beauty of biological processes and at the same time deeply convinced that these could be explained in terms of chemistry and physics. (Monod's philosophical work, *Chance and Necessity*, Judson's *The Eighth Day of Creation*, and *A Tribute to Jacques Monod*[5] give the flavor of the man and his work. François Jacob's fascinating autobiography[6] also provides insight into the philosophy of these men and their remarkable environment at the Pasteur.)

Jacob and Monod went on, as I have said, to use the bacterial mating system and β-galactosidase in their astonishing tour de force that elucidated the genetics and biochemistry of the regulation of gene expression during the next year. Back in Cambridge, I mentioned the PaJaMo results apparently without sufficient zeal, and their significance went unheeded. Francis, in fact, does not recall hearing about the experiment until much later that year when Monod visited him in Cambridge. And even then his imagination was not ignited; his reaction was one of puzzlement, a sense that it just did not fit. It is true that PaJaMo was still a rough experiment, lacking the refinement and precision given it by later work. There remained the possibility that the failure of recipient bacteria to show a delay in the onset of synthesis of β-galactosidase—an expected short period during which the cells would be constructing new ribosomes—was due to imprecision in measurement.

Thinking back now, I realize that I knew little about the ways of bacteria and lacked the conviction that PaJaMo was as probing and potent as it turned out to be. Two years later, the explanation—the existence of messenger RNA—turned out to be so simple and beautiful that I have wondered if I might not have hit upon it myself in time. Not a chance. It took two years of ripening and the combined intellects of Francis and Sydney and François Jacob and Jim Watson to penetrate to the truth.

Francis said later that the failure to find the messenger much earlier had been the one great howler in molecular biology. "The only thing one is thankful for is that it wasn't all done by someone, as it were, outside the magic circle because we would all have looked so silly. As it was, nobody realized just how silly we were."[7]

A chief clue had been lying around loose since 1953. During that year, Alfred Hershey, at Cold Spring Harbor, noticed that immediately after bacteriophages injected their DNA into their victims, a small amount of RNA was made very rapidly in the bacteria. Because it was made very fast and in very small quantity in response to the entry of phage DNA, it seemed clearly not to be ribosomal RNA. Hershey did not see what to do with the finding, so he published it and forgot about it.

The bacteriophage system is an obvious analog to the β-galactosidase mating system. The viruses attach themselves to bacteria and inject their DNA into the cells; within a few minutes, the bacterial cell's machinery begins to turn out a hundred or so new viruses. The viral DNA thus programs the bacterium to make new viral protein and DNA, just as bacterial donor DNA entering the recipient cell during conjugation programs the synthesis of an enzyme.

Three years after Hershey's experiment, Elliot Volkin and Lazarus Astrachan at the Oak Ridge National Laboratory in Tennessee made similar observations, adding the enticing morsel that the base composition of the small, rapidly made RNA resembled that of the infecting phage's DNA. It appeared that the phage was causing the bacterium to make a new RNA that resembled the phage's own genetic material. Volkin and Astrachan, like Hershey, had nowhere to plug in the finding, so they, too, put it on the shelf.

François Jacob and Sydney Brenner had had extensive experience with bacteriophages. They knew that the sole function of these viruses was to convert bacterial cells to the manufacture of virus protein, and they were familiar with the PaJaMo results. Those insights merged with the synthetic imagination of Francis Crick one April afternoon in 1960 to disinter the Hershey-Astrachan-Volkin results and create the messenger RNA hypothesis. In an inspired leap, Jacob, Brenner, and Crick postulated that this rapidly made and rapidly destroyed RNA was carrying a message from phage DNA to bacterial ribosomes. Could it be that in the PaJaMo system the entering β-galactosidase gene immediately caused just such an RNA to be made? This might be the long-sought template RNA that could become attached to the recipient's ribosomes and thereby provide the information for the lining up of amino acids in the proper sequence to make the enzyme β-galactosidase. The ribosomes were suddenly seen as a neutral factory, simply waiting for the information upon which they could make new protein. In the PaJaMo experiment, only the information had to be made, not the whole factory.

The messenger hypothesis required that the intermediary information-carrying RNA, the messenger, be unstable—in other words, be rapidly made when required by an expressible gene and as rapidly destroyed after use. Given this property, the messenger RNA needed to exist only in minute quantity, accounting for the fact that it had never been detected in ribosomes.

At this critical moment, Pardee, back at Berkeley, performed a neat experiment that demonstrated the instability of the postulated messenger RNA in another way. Essentially, he showed that the destruction of DNA in bacteria was accompanied by a rapid cessation of protein synthesis. Thus, gene action was indeed dynamic, presumably continuously making and unmaking information for protein synthesis.[8]

The messenger RNA hypothesis was precisely on the mark. Within months, Brenner and Jacob had solicited the help of Matthew Meselson at Caltech, an expert in the physical characterization of RNA and DNA, to do a sharp test of the theory. Using the phage system, they found that the putative template RNA—the RNA that is rapidly made in phage-infected bacteria—does indeed bind to the ribosomes of bacteria.[9] At about the same time, Jim Watson and his group at Harvard were uncovering similar evidence for an unstable RNA in normal, uninfected bacteria that was different from ribosomal or transfer RNA.[10] These were the first direct demonstrations that the postulated informational RNA had certain properties expected of a template. In the coming few years, proof of the validity of the concept came in rapidly, establishing messenger RNA as an essential component of the machinery of protein synthesis.

Messenger RNA tied up the protein synthesis package. Many details have since been added to the wondrous mechanism. But by 1958 we knew in broad outline how the system was energized, how genetic information in DNA programmed the ribosome, and how the cell, by means of transfer RNA, was able to put energized amino acids into proper order.

The hope that had brought me to Cambridge was, as I have said, that tRNA might help reveal the genetic code. In one sense it did, when much later the complete structure of tRNA molecules was worked out to reveal a portion called the anticodon. Messenger RNA, it turns out, contains triplet sequences of bases called codons, which unambiguously specify the order of each of the twenty amino acids to be arrayed along its length. Transfer RNAs each contain an anticodon that matches up with the codon by complementary base pairing, thus accomplishing the adaptor function. But the code was cracked much sooner, in 1960, and by an entirely unexpected means. Using a broken-cell protein-synthesizing system from bacteria, Marshall Nirenberg and J. H. Matthaei at the National Cancer Institute showed that by adding man-made messengers—unnatural RNAs composed of, for example, a single base repeating over and over—the system would make an artificial protein made up of a repeating single amino acid.[11] In other words, the natural machinery of biosynthesis would translate *any* message, natural or artificial, into a polypeptide chain, a protein, natural or artificial. By varing the nucleotide content of the artificial messenger, one could require ribosomes and tRNA to make polypeptides of varing composition. In this way, the sets of nucleotides that code for each of the twenty amino acids were eventually disclosed. I should note, in closing this account of the triumphs in dissecting the protein synthesis machinery, that the bacterial system Nirenberg used was one first worked out by Paul Zamecnik and his young associate Marvin Lamborg in 1958–59 at the Huntington.[12]

I left Cambridge in the summer of 1958 with mixed feelings. I had greatly enjoyed the professional companionship, the education, and the hobnobbing with the nobility in the court of molecular biology at the peak of its greatest glory. I had lived with my family in the exquisite little hamlet of Grantchester, two miles up the Cam from Cambridge, in a charming thatch-roof house known as Byron's Lodge. I had been made an honorary member of the local pub's dart team. I had enjoyed the privileges of Francis's

Gonville and Caius College. I had spent many pleasant evenings with Francis and Odile and their friends at their delightful home at Portugal Place, with its golden helix over the front door, eating well, talking, and singing our songs. (There was one memorable evening when we celebrated Fred Sanger's Nobel Prize. At the height of the festivities, someone set a rocket off from the roof of the house. Its ill-planned trajectory caused it to lodge, still smoldering, in the tower of the nearby church, and this necessitated calling the fire brigade.)

But there was a twinge of melancholy, too. For me, an idyllic decade was drawing to a close. My association with the MRC group had convinced me that by temperament I was ill suited to continue to compete in the game. Our horizons had been expanded by the pioneers, and there was a lot of obvious things to do: proving colinearity, working out the biochemistry of DNA and RNA replication, pursuing mechanical details of protein synthesis, teasing out the genetic code. These were easily visualizable goals, and their achievement looked to be inevitable. They would be pursued frenetically, in an excessively competitive atmosphere, by bright young scientists who had been attracted by the triumphs of molecular biology. Those goals did not smack sufficiently of mystery, drama, and adventure to suit me.

7

❖ ❖ ❖

Entr'acte

My Cambridge sojourn and return to the Huntington prompts a reflection on the qualities of the two men who loomed so large in the first decade of my professional life. Paul Zamecnik and Francis Crick are about as different in personality and style as any two scientists I have known. Paul is as pure an experimentalist as is likely to be found in biology today; Francis, as pure a theorist. Paul came to biology from medicine; Francis, from physics. Paul is taciturn, noncommittal, even enigmatic; Francis is explicit, voluble, and opinionated. Paul is politically and socially conservative; Francis, liberal and iconoclastic. Paul is patiently persistent in the pursuit of his own experimental goals; Francis invades and tackles the problems of others. Similarities? Both have an intuitive sense of where the pay dirt lies and of how to grasp opportunity and exploit advantage; both know better than most what they do not know. Both are tolerant of scientific aggressors, effective in competition without appearing to be competitors, and both are warm, humorous, considerate, pleasant company.

Paul's remarkable successes in dissecting the mechanism of protein synthesis—to say nothing of his continuing productivity in other important areas of research since then—

deserve much greater recognition than they have received. Just as Francis in 1947 saw biology in terms of molecular structure, so Paul at about the same time saw the synthesis of living substance in terms of energy. Each made imaginative use of the available technologies: Francis, of X-ray diffraction; Paul, of radioactive isotopes, early techniques for analyzing protein composition, and the ultracentrifuge. Francis's success with Jim Watson in picturing the structure of DNA and his vision and eloquence in formulating the grand objectives of molecular biology were inspirational. Pauls' decade of continual creativity in protein synthesis was a sweating, earthbound, innovative struggle that laid bare many of the intricacies of any incredibly complex machine.

Paul joined the Huntington in 1936, became its director in 1956, and continued to follow Joseph Aub's visionary commitment to fundamental studies of normal growth as a basis for understanding cancer. He says he was propelled to go into research by early ponderings of the possible existence of chemical elements that controlled the growth of cells, aberrations of which might result in the uncontrolled proliferation of cancer cells. During his career at Harvard, as leader of the Huntington group, as professor of oncologic medicine, and as a mover in policymaking at the Massachusetts General Hospital, particularly as it related to keeping the hospital strong in basic research, his quiet sincerity was highly effective and widely appreciated.

Under Paul's guidance, we at the Huntington learned by osmosis and example and by doing things ourselves. While "out there" in the scientific world our colleagues seemed to be in cutthroat competition, in the Huntington there was a fraternal ease. Most of us—some dozen scientists, an equal number of technicians, and our families—remember ski trips in winter, beach jaunts in summer, and parties with much originality in the form of skits, caricatures, and songs. Work conferences were lively and relaxed, lit up by the occasional practical joke. In those days, we always seemed to be in front of the field without feeling that we were even trying very hard.

Perhaps the most lasting impression I have of the Huntington is as a free and open place where the creative person could find self-expression and would stand or fall on his or her own talents, without coddling or coercion. But for me it was time to move on. The Huntington had been a supportive scientific base for a decade. It was hard to envision any other home, but my reputation was too closely tied to Paul, and I needed to plunge into uncharted territory on my own.

I did a final set of necessary experiments on transfer RNA to show that the whole tRNA molecule accompanied the amino acid into the ribosome during protein synthesis.[1] But that last year and a half at the Huntington was taken up largely with writing two major papers of more general interest. One was an extensive review of the scientific literature on protein synthesis in cell-free systems;[2] the other was a more popular account of genetic coding and the role of DNA and RNA in protein synthesis.[3] The review was satisfying to me as a scholarly job by someone who did not consider himself very scholarly. The *Scientific American* article pleased me because as far as I know it was the first popular account of the subject by a scientist. *Scientific American* accepted my own illustrations, and the article proved to be one of the magazine's most widely reprinted. Later, I put substantial effort into the writing of an introductory chapter on coding, information transfer, and protein synthesis for *The Metabolic Basis of Inherited Disease*, which through two subsequent editions again exposed a wider audience to the ways and achievements of molecular biology.[4] These experiences were the beginnings of a greater immersion in writing about biomedical science for the general public.

It was dawning on me that the freewheeling, carefree

career I had chosen and enjoyed for nearly twelve years was carrying me inexorably toward *professing.* I was offered a job as chairman of the Department of Physiology at the University of Chicago, and it brought me up short. Did I, at age thirty-nine, want to plunge into teaching, part-time research, administering, coddling graduate students, planning exams, correcting papers? I had not given the future a thought, and now, confronting it, did not much like what I saw. My experience differed from the usual pattern of academic life—getting a Ph.D. degree, doing two or three years of postdoctoral work, and then starting the climb up the university academic ladder en route to lifetime tenure as a professor. I completed an M.D. and went into full-time, open-ended research for a dozen years, at the end of which I attained the lowly rank of assistant professor of medicine at Harvard. I did have a short stint in one other post: from 1954 to 1957, I served as the executive secretary of the MGH's Committee on Research. I valued that experience because it gave me broad contact with other research activities at the hospital and had been a relatively easy-to-take diversion from full-time research.

The decade of the 1960s was one of much upheaval and renewal for me. My seventeen-year marriage ended in divorce, followed by remarriage, to Olley Jones Robbins. Seven years at Harvard Medical School and three at Dartmouth Medical School brought me the satisfactions of teaching and new friendships and the continuing pleasures and growing frustrations of research, its conduct increasingly in the hands of younger colleagues. The Worcester Foundation continued to lurk in the wings, prodding my conscience and, eventually, attracting me. Nixon and Vietnam spilled into our lives and darkened them. One of my daughters was a victim of that bewildered, drugged decade, ending her life in suicide.

Olley Jones Hoagland, 1962 *(Lynn Sanford, Essex, Conn.)*

In 1959, I received an offer of an associate professorship at Harvard Medical School's Department of Bacteriology and Immunology (now Microbiology and Molecular Genetics). I could still devote most of my time to research of my own choosing, do a limited amount of teaching, remain in an exciting intellectual environment, and have tenure. The offer was too good to turn down. Willy-nilly, I *was* going to profess!

Bernard Davis, chairman of the department, was good enough to put in my hands the more challenging lectures on the emerging knowledge of regulatory mechanisms in bacteria. I found the task difficult, scholarly, and immensely enjoyable. Bernie Davis was an exacting and sometimes exasperating boss who attended my lectures, took notes, and criticized me mercilessly afterward. He earned my admiration by his wide-ranging scholarship in microbiology and molecular biology and by his exceptional qualities as a scientist and teacher. He has been a controversial figure in American biology because of his bluntly aggressive confrontation of those who would qualify and politicize the search for truth. He has defended human genetic and behavioral research against colleagues who sought to halt it because of imagined or potential abuses. He was widely vilified but ultimately vindicated for his stance against Harvard Medical School's policy of sacrificing academic standards on the altar of affirmative action. And he played a key role in defusing the explosive national debate on recombinant DNA research by introducing into it rationality and a sound base of microbiological knowledge. His zeal, intellectual courage, and intolerance of hypocrisy, coupled with a political and social insensitivity, have kept him at the center of many issues relating to science, health, and public policy and at the same time have alienated some of his colleagues. Our friendship has had its peaks and valleys but has been sustained by mutual respect, shared values, and his excep-

Bernard Davis, ca. 1980 *(courtesy Bernard Davis)*

tional talent for always being interesting—and for telling good jokes. His collected essays, published as *Storm over Biology*, give the flavor of the man in clear and provocative prose.[5]

To Bernie's disappointment, I never became a microbiologist. Perhaps the dramatic changes in my personal life coinciding with my entry into the department made me want to stay in familiar territory. But I was deeply attracted to animal cells. The one research problem that intrigued me throughout much of my career was an attempt to explain the phenomenon of regeneration. I was about to spend nearly ten years struggling with the problem, only to emerge chastened—aware of my own fallibility and impressed with the creative imagination of the younger colleagues who came to work with me.

The liver of the humble rat had served well as the source of many of the key insights into the mechanisms of protein synthesis. That it could be induced to regenerate at will had offered biologists some exciting prospects for fathoming the complexities of the *regulation* of synthesis of large molecules—protein, RNA, and DNA. Regeneration is truly an awesome phenomenon, in which the reawakening of embryonic "memory" is manifested in the orchestrated proliferation of cells to sculpture a whole new, highly complex organ. Equally remarkable is the regenerating organ's "knowledge" of when to stop. Bacteria, higher plants, and cancerous animal cells divide and grow indefinitely at rates largely determined by the availability of nutrients in their environment. Normal cells of most adult animal organs, on the other hand, live for long periods, some as long as the animal itself, without dividing. Other animal cells, such as those of skin, intestinal mucosa, blood, and liver, reproduce at varying speeds, but always in such a way that cell

Nancy Bucher, 1953 *(courtesy Nancy Bucher)*

death balances cell birth, the total number of cells remaining fixed. This exquisite homeostatic control extends to events *within* cells as well, where cellular constituents, like proteins, RNA, and fats, "turn over" constantly—in other words, are destroyed and rebuilt at equal rates.

I mentioned earlier my brief flirtation with the regeneration of tadpole tails in connection with the beryllium studies. Later, at the Huntington, I was exposed to the fascinating investigations of Nancy Bucher, who used parabiotic rats—surgically constructed Siamese twins—to study biochemical events during regeneration. This experimental system made it possible to remove part of the liver of one partner and study the effects of the residual fragment's regeneration on the liver of the unoperated partner. The existence of circulating factors that influence cell proliferation could thereby be detected.

Ten years previously, Paul Zamecnik and researchers in Sweden had injected radioactive amino acids into rats whose livers were regenerating and into normal controls. Both groups concluded that regeneration was accompanied by a substantial increase in the rate of protein synthesis. But in the time since, as more and more workers focused on the intimate details of protein synthesis, little had been done to follow up on those early studies, which, as we will see, proved to be misleading.

During the 1950s, bacteria provided a wealth of information about the regulation of the synthesis of *specific* proteins and about the regulation of bulk protein content in cells. When these organisms grow at a given rate, determined by the availability of nutrients in their environment, the relative amount of protein synthesis machinery—notably ribosomes—and of total protein per cell is predictably proportional to that growth rate. In general, the faster the growth, the higher the content of both. During the early phase of transition from slower to faster growth, the amount of machinery increases first, followed by an increase in total protein. This response is not to be confused with the result of the PaJaMo experiment, in which a few ribosomes are very rapidly programmed with a new messenger RNA upon introduction of a foreign gene into a cell. It would appear, then, that these relatively simple organisms use their synthetic apparatus at top capacity at any given steady-state growth rate. (In the transition to slower growth, machinery construction ceases and the components gradually disappear as the number of cells increases by continuing cell division.)

The data were sparser for animal cells. In general, like bacterial cells, animal cells growing at their different characteristic steady-state rates showed a rough positive correlation between growth rate and ribosome content. But what might occur upon induction of rapid growth—as in regeneration—in a dormant population of adult, nongrowing animal cells, such as those of the liver, was unknown.

In launching a new research program at Harvard, I determined to try to use regeneration as an entry to understanding the control of protein synthesis. As a long shot, I postulated that there might exist in normal, adult, nongrowing cells a factor or factors responsible for holding protein synthesis in check. Such a factor might initially be detectable by its disappearance at the beginning of regeneration. My hope was that it might manifest itself as a lower intrinsic rate of synthesis in protein synthesis machinery isolated from normal liver as compared with the rate in machinery from regenerating liver. If this were so, I had a chance of finding the factor.

Unlike those in bacteria, however, most of the ribosomes in liver cells are bound to a highly organized membrane network called the endoplasmic reticulum. When cells are broken open, the ribosome-membrane complexes are isolated in the ultracentrifuge as particles called microsomes.

In a search for elements that might control protein synthesis on ribosomes, it seemed important to try to preserve their natural relation to membranes. I therefore carefully extracted microsomes from normal and regenerating liver and incubated them with the appropriate components for protein synthesis. I found that the latter were indeed substantially more active in synthesizing protein than the former. This seemed to confirm earlier claims of higher activity in regenerating liver in the living animal.

The lesser activity of normal, nonregenerating microsomes appeared to be attributable to the membrane material associated with their indigenous ribosomes—in other words, that material inhibited protein synthesis, as I had hoped it might. I first found this by mixing microsomes from normal and regenerating livers and observing that the synthetic activity of the mixture was about the same as an equal amount of normal microsomes. I was then able to extract from broken-up microsomal membranes material that inhibited protein synthesis in fresh active preparations of regenerating microsomes.[6]

In 1963, Oscar Scornik, a postdoctoral student from Argentina, joined my group. Oscar is a brilliant experimentalist: focused, perceptive, original, and profoundly skeptical of claims he has not himself verified. He took up the regenerating-liver system and, exploring its properties with characteristic care, revealed that the inhibitory property of normal liver microsomes that I had found was essentially an artifact. These crude microsomes, it turned out, also contained lysosomes—cell bodies harboring enzymes capable, when released, of breaking down many cellular components, including ATP, RNA, and proteins. It appeared that the lysosomes of normal liver were more fragile than those of regenerating liver when removed from the liver and incubated by our methods. This caused them to release their degradative enzymes, thus inhibiting protein synthesis by destroying essential elements of the machinery.[7]

Oscar Scornik, 1988 *(courtesy Oscar Scornik)*

(Other nonlysosome inhibitory factors in microsomes that intrigued us for several years were later shown—by another postdoc in my lab, Roger Quincey—to be due to additional kinds of degradative enzymes similarly released in the act of preparing the active fractions.[8])

Breaking open cells to study their inner workings produces great devastation. Cells are highly integrated, complex worlds of intricate machinery in which proximities and relations of components are essential to proper function. It has always struck biochemists as truly remarkable that they have been able to derive such wondrous detail from the chaos produced by breaking open cells. It testifies to the impressive success of the analytical approach—the reductionist approach—in which the study of parts has led to an amazing revelation of the whole.

In 1966, Oscar returned to Argentina. Dissatisfied with and increasingly suspicious of all earlier studies of protein synthesis in regenerating liver, Oscar decided to get out of the foliage and back down to the roots. He undertook a thorough investigation of what was happening to liver protein content in the early stages of regeneration in living animals. In the ensuing three years, this experimental tour de force invalidated most of the previous work in the field and opened the door to a major new insight into growth control in animal cells.

In cells of the adult, nongrowing liver, there is, as was mentioned earlier, a high rate of turnover: rapid protein synthesis *and* rapid protein degradation. Synthesis and degradation exactly balance one another, so the liver does not change in size. Most of the liver's protein substance is involved in this process, which appears to serve the purpose of storing amino acids obtained from food as protein and releasing them from protein during periods of fasting.

During regeneration, Oscar found, the number of ribosomes and the rate of protein synthesis changes very little, but *protein breakdown is markedly suppressed*, thereby allowing the overall protein content of liver cells to increase rapidly. Indeed, the restoration of most of the mass of the liver (a threefold increase in total mass in a three-day period) can be largely accounted for by suppression of protein degradation. The process is strikingly different from that observed in bacteria, in which there is very little turnover of protein and a new growth demand is accompanied by the construction of additional protein-making machinery.

Oscar's studies point out the obstacles to getting unambiguous results from studies of protein synthesis using radioactive isotopes in whole animals. The false conclusion of previous studies that indicated increased protein synthesis during regeneration stems from a tricky feature of the use of isotopically labeled compounds to study the metabolism of the same compounds being produced in the body. As regeneration gets under way, protein breakdown is drastically curtailed, thus slowing the internal production of (nonradioactive) amino acids. Injected radioactive amino acids are, as a consequence, substantially less "diluted" with nonradioactive amino acids; amino acids entering protein are therefore "hotter," giving the erroneous impression that protein synthesis has increased.[9]

Oscar's thoroughly documented work, much of it done in collaboration with his wife, Violeta Botbol, has clearly established, that, in the restoration of liver mass during regeneration, the cessation of degradation of liver protein is far more important than new synthesis.*

---

* Liver and other mammalian organs in the normal, nongrowing state are full of active lysosomes—so-called autophagic vacuoles—that are engaged in scavenging and degrading a variety of cell constituents. Lysosomes appear to reduce their degradations during stimulation of growth. It should not be inferred, however, that the level of protein breakdown characteristic of the nongrowing state is a consequence of the lysosomal fragility that Oscar found earlier. There is no indication that the autophagic vacuoles found in nongrowing cells are more fragile in the living cell. Indeed, their release of enzymes in the living cell would be suicidal. It is only when these enlarged lysosomes are removed from cells as microsomes that they manifest their greater instability—again, a hazard of work with broken cells.

Later on, after I moved to Dartmouth Medical School, Oscar accepted my offer of a faculty position. He is now a professor of biochemistry there. He and Violeta have continued to build a solid base of knowledge of protein metabolism in liver and to offer new insights into growth control.

In retrospect, my attempts to fathom protein synthesis control mechanisms in a cell-free system from an organ as complex as mammalian liver appear premature and ingenuously optimistic. Even today, twenty-seven years later, the regulatory aspects of the process remain a mystery. I was chastened by the experience. My attempt to launch a new investigation from an admittedly shaky base caused some wilting of my confidence. Oscar had the requisite skepticism, imagination, and determination to construct a new base and begin to build upon it. Our best students leave us in the dust, as indeed they should.

While the struggle to fathom regeneration claimed much of our attention from 1961 to 1967, others among a fine group of associates in my laboratories at Harvard and later at Dartmouth contributed to the development of knowledge about protein synthesis, about RNA synthesis, and about the structure, function, and biosynthesis of ribosomes and ribosome-messenger RNA complexes (polyribosomes) in bacterial and animal cells.[10]

Elmer Pfefferkorn, a member of our Harvard department, was then emerging as a highly original and unorthodox researcher and a brilliant teacher of virology. In 1966, he announced that he was leaving us for a post at Dartmouth Medical School. Although one of the country's oldest and most reputable schools, it offered two years only, after which students transferred to four-year schools to complete their

Carleton Chapman, ca. 1970 *(Bachrach)*

training. At the time, Dartmouth Medical School was dusting itself off in the aftermath of a nasty academic imbroglio that had left its faculty embittered, chastened, and eager to get back on an even keel. The trouble arose from a protracted dispute over the introduction of a major program of graduate studies in molecular biology into a conventional hospital clinic–medical school setting. The program, desired by most of the basic-science faculty, was opposed by others who saw the program as the tail of avant-garde research wagging the dog of medical tradition. The controversy peaked in 1965 with the simultaneous resignation of the chairmen of three departments—biochemistry, microbiology, and anatomy. By 1966, two of the chairs had been filled, but that in biochemistry remained open. Carleton Chapman, a scholar and a dynamic and highly respected physician, had been appointed dean and was initiating a move to make Dartmouth Medical School a four-year, degree-granting institution.

I was restless. My interest in teaching had grown, but I had been distressed, both as a former medical student and now as a teacher, by the vast number of formal, often irrelevant, lectures given by a faculty more interested in research than in teaching, and by the widespread lack of interdepartmental coordination of teaching. I was unexpectedly developing some fervor for educational reform. Harvard's size, conservatism, and complacency were daunting to the missionary. Dartmouth seemed to offer a chance: relatively small size (one hundred students compared with Harvard's five hundred), the decks cleared after the recent storm, an empty department chair, a sympathetic dean, and the shift to a four-year program.

Olley and I visited and were beguiled by the school, its people, and the beauty of the countryside. We were encouraged particularly by the impressive competence of Carl Chapman, in whom we saw the promise of friendship as well. We had always lived in cities, and the rural setting suddenly looked enticing. It meant Olley's leaving a job she loved, as head of Harvard University Press's display room in Cambridge, but we sensed adventure and headed north.

The move to Dartmouth was the fourth Olley and I had made in six years, and we hoped it would be our last. In the next eighteen years, we were to change our home six more times. The Dartmouth sojourn proved to be anything but a settling down to the contemplative life of research and teaching.

I was successful in substantially revising the biochemistry curriculum. We eliminated a large number of formal lectures and got the faculty to work more closely with small groups of students in problem-solving sessions. This required a doubling of teaching time, which the faculty accepted reluctantly at first but with growing satisfaction as it became clear that the students liked the new format. Indeed, the students were more actively involved and had more time to work through problems related to their interests. But the message was purely local and of short duration. Years later, I learned that the time we had freed up by teaching biochemistry more efficiently was quickly grabbed by other departments to extend their territorial claims on the beleaguered student. Again, interdepartmental coordination remained a dream. In all my experience with academic inertia and rigidity, medical school curricula stand out in their steadfast resistance to innovation.

I had hoped to strengthen the biochemistry department by recruiting new talent in molecular biology. But the confrontation between my impatience and the department's intransigent conservatism generated more heat than light. In frustration, I asked Carl Chapman to help convene an outside committee to evaluate my assessment of the department's needs. A distinguished group of colleagues made a site visit. Their attention was largely diverted from my proposals, however, by an obsession with the notion that the medical school Department of Biochemistry should merge with the neighboring college Department of Biology. This idea had some merit, but it had been proposed and rejected several times in the past and had no appeal for me, anyway, since there was no hope of expanding the commitment of either college or medical school to acquiring a more talented faculty. Never in my career has an idea of my own exploded so dramatically in my face!

I naïvely waded into another swamp by advocating a departmental merger within the medical school. The departments of biochemistry and microbiology had common research orientations and teaching objectives, yet each taught medical students as though it did not know the other department existed. The chairman of microbiology wanted out, and a merger of the two small departments seemed to me a unique opportunity to encourage interdisciplinary teaching and research. The chairman of microbiology and I, with Chapman's encouragement, broached the idea to the respective faculties. I should have known better. Again, I emerged a trouble-making interloper in the gentle hills of New Hampshire, but a bit wiser about how to sell ideas.

The Worcester Foundation, of course, would not go away. In the fall of 1965, Gregory Pincus came to visit me with purpose written on his face. A decade had passed since his immortality had been assured by his commitment to the birth control pill and its powerful social implications, and he would have only two more years until his death. My relationship with Pincus before this visit had been that of a young man with a benevolent, kindly uncle. As my father's lifelong friend, Pincus had always been to me gentle, attentive, amusing, and understanding. While some of his colleagues saw an aggressive scientific entrepreneur, I saw a sort of family Einstein: a head full of wisdom, with penetrating kindly eyes and topped by a surprised mass of black hair.

As I related in chapter 1, I was starting medical school when my father and Pincus were launching the Worcester Foundation. Now, here was my father's partner as a fellow professional with a problem on his mind. His visit intrigued and honored me. He talked about the Foundation and its future. Both he and my father were approaching retirement age, and the continuity of their joint enterprise was very much on his mind. Pincus was away from the institution a lot, and my father was growing restive as the resident administrator of a scientific enterprise inspired in the main by Pincus's interests and successes.

Specifically, Pincus was troubled by the Foundation's lack of any representation in molecular biology. It was a matter about which I had expressed concern to my father on more than one occasion. Pincus had a solution: *I* should move to the Foundation and establish my research program there. Without explicitly stating it, he clearly saw this as a solution to the wider problem of leadership continuity. He said he had not discussed his visit with my father, but I knew that he and my father had not been unmindful of the possibility that I might eventually be persuaded to join the Foundation.

I am convinced that neither of these men was then aware of the complexity of the problem of ensuring the Foundation's future. By 1965, the institution had grown to the largest size it was ever to attain, driven by the two men's ambition and a rapidly expanding base of federal research support. Neither of them seemed inclined to limit the institution's expansion as long as money was available. The absence of any significant endowment, however, made it exceptionally vulnerable. And while I did not say so to Pincus, I was not overly impressed by the quality of some of the science I had seen at the Foundation. On several occasions, while serving on national panels of the National Institutes of Health and the American Cancer Society, I had evaluated grant applications submitted by Foundation scientists; some had ben singularly weak. I had a sense that there was too casual an attitude toward faculty quality and a laxity in administrative control. By 1965, my father and Pincus were riding an obstreperous twenty-two-year-old bronco upon whose back others would not be too willing to climb in their place.

While I was moved by Pincus's concern, I had little inclination to become involved. It seemed to me both he and my father were vigorous and had adequate time for planning. Whoever eventually succeeded these two charismatic men should be accomplished in the fields of the majority of scientists there and should not be the son of one of them.

That conversation, sadly, was the last I ever had with Pincus. Three years later, one year after my move to Dartmouth, I was again approached, this time by a search committee of the Foundation's board of trustees asking me to consider becoming the director. Much had happened in the meantime: Pincus had died and my father retired. Federal funding for research was beginning to plateau after two decades of exuberant growth. Some of the Foundation's more able scientists, those who had job offers, were leaving. A nonscientist manager who had earlier been appointed as an administrative assistant was acting as director. He was having a tough time holding the operation together as he coped with growing financial and grants management problems and with a sharp collapse of the morale of scientists and staff. I declined the offer, agreed to look over the dossiers of some thirty of its scientists in a kind of advisory role—and felt troubled.

On rare occasions when my father and I got together, he continued to speak, with a shade too much bravado, of the Foundation's bright future. Able candidates, he reported, were considering the directorship. His apparent optimism was partly responsible for my not being fully aware of how bad things really were. He refused to indicate to me any personal wish that I might change my mind, insisting that I continue my research. He evinced much nostalgia for his own, simpler days of research, though almost everyone agreed that, until recently, he had immensely enjoyed his years of managing the Worcester Foundation.

Between my frustrations at Dartmouth and the Worcester Foundation's troubles, I was in a quandary. The balance shifted during a visit of a colleague, Donald Wallach, whom I had been promoting against faculty opposition for a post in the Department of Biochemistry. Sitting at home one evening, he had asked me about the Foundation, and I told him of its troubles. He wondered, at what in retrospect was that critical moment when one needed to be pushed off the fence, why the situation did not offer an almost ideal opportunity to create a new program of my own designing. Well, why not? My real interests lay in research. I had lost my shaky affection for academia.

At about this same time, the British science journal *Nature* published an editorial on the Worcester Foundation, the first part of which read like an obituary. It went on to say,

> The immediate need is the appointment of somebody who will be responsible for the balance of the research programme, and it is more than likely that a man of distinction great enough to maintain the momentum of the past decades is likely also to bring about substantial changes in the pattern of work. Uncertainty about the future of Federal support for scientific research also makes the future harder to predict, although the Worcester Foundation seems to be a good deal more cheerful on this score than a great many university departments where federal research grants make a smaller contribution to the operating budget. . . .
>
> To outsiders, the foundation is every bit as remarkable as its supporters say. It is a striking embodiment of the old puritan virtues endemic in the hinterland of Boston. . . . Senior members of the research staff emphasize the importance of insisting that the work carried out should be of the highest quality (which in turn emphasizes the need that somebody of distinction should quickly be appointed to take charge of the scientific programme). The question of whether all this can be accomplished without continued growth remains to be resolved and this, of course, is not an open question—much will depend on whether NIH really does have extra money to spend in the years ahead.
>
> The geographical isolation of the foundation, in many

> ways an asset in the past twenty years, could also become a disadvantage if there were too rapid a turnover among the distinguished people who at present serve as magnets for recruits. It is thus more vulnerable than many universities, especially now that university salaries can be extremely tempting. It may yet, however, become an interesting demonstration of how it is possible to build from scratch a productive ivory tower in the wide open country.[11]

Any further consideration of the pros and cons was moot. The more Olley and I learned about the plight of the Foundation, the more we realized that we had no choice. The institution was clearly going to founder without leadership.

The sigh of relief from Roy Greep, chairman of the search committee, when I called him was clearly audible. His quick assurance that it would not be difficult to get the other members' approval implied the job was mine. He was shedding a heavy burden right there on the telephone. I rang off with the sinking feeling that all those who had turned down the job had some kind of sense that I lacked.

8

❖ ❖ ❖

Two Traditions

I arrived at the Worcester Foundation in the summer of 1970 excited, worried, ill prepared, and firmly convinced that the institution must be moved from its traditional emphasis on physiology and biochemistry to the conceptual and experimental ways of molecular and cell biology. Cell biology can be described as the extension of molecular biology's use of genetics, physics, and chemistry to probe the structure and function of animal cells in culture. The growing and manipulating of bacterial cells had been relatively easy compared with the handling of animal cells. The technology for the culture of animal cells—tumor cells, connective-tissue cells, and the separated cells of certain organs—began to mature in the 1960s. The progress made it possible to handle animal and bacterial cells in similar ways, including control of their environment and the production and selection of mutants. This focus on animal cells was the next logical thrust into a deeper layer of complexity, reflecting the optimistic view that success with bacteria and biochemical dissection provided a base for understanding the cells of higher organisms.

The phenomenal success of biochemistry and molecular biology derived in part from an analytical, or reductionist,

The Worcester Foundation, 1988 *(Sky Hawk Photo Service)*

assumption. By focusing on the physical and chemical properties of such large polymer molecules as proteins, DNA, and RNA—molecules considered to be capable of carrying genetic information in their sequences—science had unearthed surprisingly complex function residing in structure. For example, the revelation of the structure of DNA had also revealed its mode of replication, the first step in creating a new living entity. Enzymes—protein molecules carrying out life's functions—were able to do much more than simply catalyze essential chemical reactions. As simple polymers of amino acids, they naturally folded into specific complex conformations that displayed remarkably sophisticated properties. For instance, enzymes were found to be able to regulate their own rate of performance under the influence of metabolic products that bore no resemblance to the substances they acted upon. They were cybernetic, or feedback control, systems of astonishing versatility, yet entirely explicable as chemical entities. And the elucidation of the nucleic acid–protein translation machinery of cells had disclosed a wealth of intricate mechanisms whereby energy is consumed to convert linear molecular information into the three-dimensional substance of living cells.

The study of "simple" bacteria had revealed intricate details of how they regulated, through the interplay of large molecules, their internal composition, their nutrition, and their ability to adapt to environmental vicissitudes. New light on genetic variation revealed, as I noted earlier, that bacteria were covoyagers with all other life forms on the evolutionary journey, subject to the same mechanisms of self-assembly and the same forces of mutational change and selection.

Already some aspects of differentiation and the genesis of form and pattern were looking like a spontaneous interplay and assembly of versatile proteins and RNA or DNA. For example, the study of the properties and the assembly of viruses had led to some significant insights into morphogenesis—self-assembly processes—in general. All of these revelations had given scientists enormous confidence that biological systems of a much higher order of complexity could be comprehended—that we could get at whole-cell and whole-animal complexity—by studying the properties of parts, particularly big molecules. Molecular biology made more absurd than ever the mystical holistic, vitalistic view that there was something unfathomably complex and incomprehensible about living systems that no analytical attack could ever explain. A living organism was indeed much more than the sum of its parts, but so was a protein. Lowly molecules were proving to be loaded with the kinds of "knowledge" that could make incredibly complex patterns and events happen.

The extraordinary successes of these new ways of looking

at biological problems produced an elite core of young scientists in the United States, Europe, and Japan who were trained in molecular genetics, biophysics, and cell biology. They were poised in 1970 to put real muscle and mind into biology's hottest problems: cell growth control, mechanisms of differentiation and carcinogenesis, host-virus relations, immune mechanisms, molecular mechanisms of hormone action, fertilization, and some aspects of nerve and brain function. I wanted to build this profound new competence and optimism into the fiber of the Worcester Foundation by assembling a new group of cell biologists imbued with the faith. I hoped, too, to strengthen ongoing Foundation activities with the new people and the new faith.

However, few of the incumbent scientists at the Foundation fit with this agenda. Gregory Pincus had been aware of this, as I related in the last chapter. They, like most of their colleagues in similar fields throughout the country, explored the reproductive process, hormone function, and the nervous system, mostly in the whole animal, using conventional techniques and concepts of physiology and biochemistry. The ways of the new cell biology, with its focus on structure-function relations on large molecules studied in simpler, graspable cellular model systems, were only beginning to penetrate these areas of research in the late 1960s.

Another practical and philosophical problem confronted me at the Foundation. While both Pincus and my father believed in and eloquently espoused the importance of free exploration, and while most of the scientists obtained their own research grants, the institution was much more attentive to practical medical and social application than I had been in my professional life up to this point. Over the years, for example, the pharmaceuticals industry had supported parts of the Foundation's research, particularly in endocrine biology. In addition, there was a growing acceptance among Foundation scientists of research supported by contracts with industry and government, directed toward relatively short-range objectives. The great success of the contraceptive pill had been both a result and a cause of this state of affairs. And the relative instability of the institution's financial infrastructure encouraged this use of product-oriented, and to me distasteful, means of science support.

The emphasis on social relevance had deep roots in my father's and in Pincus's early training and experience. They had both taken their doctoral degrees with the head of Harvard's Laboratory of General Physiology—the brilliant, belligerent, opinionated William J. Crozier. The noted behavioral psychologist B. F. Skinner was also among Crozier's young Ph.D. students. Crozier's hero had been Jacques Loeb, a leading figure of American experimental biology in the early 1900s. Loeb had made headline scientific news in 1899 with the discovery of artificial parthenogenesis, in which he induced sea urchin eggs to begin development by stimulating them with certain salts—without benefit of sperm. He also pioneered the investigation of animal behavior and made a valued contribution to biology by emphasizing the importance of applying principles of physics and chemistry in experimentation. Widely influential among his fellow professionals and educated layman, Loeb perceived science as a powerful and challenging means of controlling, improving, molding, and, indeed, engineering life for the benefit of humankind. (H. J. Muller, the discoverer of the X-ray induction of mutations and an outspoken advocate of the use of genetics to improve the human gene pool, was also strongly influenced by Loeb.) Thus, through Crozier, my father and Pincus were imbued with a zeal for both the engineering approach and, perhaps prematurely, the application of physical chemistry to biological problems. It was not, however, until the 1950s that physical chemistry was applied to biology, by way of molecular biology, with truly telling effect.

My father chose neurophysiology; Skinner moved into

animal behavior. Pincus was spurred by Crozier to apply his exceptional abilities in physiology, chemistry, and genetics to the study of mammalian reproduction, and early on Pincus came to think in terms of the control of the process. His initial focus on artificial parthenogenesis in rabbits (work for which he gained early notoriety but which later proved to be in error), on in vitro fertilization, preservation of sperm, artificial insemination, and the control of ovulation were clearly products of this background. My father's and Pincus's common interests in industrial-worker and pilot fatigue during the war, and in neural and hormonal factors in schizophrenia, also reflected the Loeb-Crozier influence. The climactic discovery of the oral contraceptive seemed an almost inevitable outcome of Pincus's fervent drive to vindicate a philosophy of research that emphasized human benefit.

The historian Philip Pauly has recently pointed out that Pincus's emphasis on biological technology was a source of criticism from some colleagues who did not regard this priority as befitting an academic scientist. In this respect, Pincus was an early practitioner of biotechnology.[1] The readiness with which he responded to the concerns of Margaret Sanger and Katherine McCormick, his involvement with the pharmaceuticals industry, and his interest in John Rock's work with infertile patients were parts of the pattern.

All of this was very different from the tradition of basic biochemistry and of the molecular and cell biological exploration of the 1940s and 1950s, in which I had been nurtured. In that world, knowledge of information transfer and the regulation of macromolecular synthesis had advanced on the wings of pure science with hardly a thought to practical application or relevance. The goal was understanding, as purely and simply as at any time in the history of biological exploration. There never would be, in my view, a shortage of those who would apply knowledge to human needs and desires once it was made available, but they were outside the fraternity. When, in the 1970s, recombinant DNA technology emerged from the pure research of molecular biology, the Loebian dream of engineering for human welfare came to play an increasingly significant role in the thinking and behavior of biologists, this time augmented by the profit motive.

However objective we may try to be, scientific trends are born not only of the state of the art but of intuition and aesthetic taste as well, and sometimes of a scientist's own personality and persuasiveness. I for one was convinced that good, solid, ivory-tower cell biology offered the sharpest knife for dissecting some of the more interesting problems of the coming decades, and I was determined that the Worcester Foundation should use it. In the broad view, it could be said that the triumphs of the era 1944–70 had been the discovery of universal structures and mechanisms relating to information transfer and macromolecular synthesis. The great upcoming task, as I saw it then, was to discover the rules governing the emergence of difference, of increasing complexity, in biological systems.

During the three or four years prior to my arrival at the Foundation, it had become apparent that steadily rising federal support for biomedical research, to which we had become accustomed in the 1950s and early 1960s, was leveling off. We were entering an era—the one in which we now find ourselves—of a waning national spirit for exploration. The greater austerity forced upon us would preclude expansion. Fortunately, the Foundation appeared to have attained a comfortable, even an optimal, size of some 35 scientists and an additional 250 support staff. It would not take too many more people to cause the inevitable fragmentation into separate research departments, and I wanted to avoid that if at all possible. A change in research emphasis, then, would require a substantial replacement of faculty and a lot of money from new sources. Some scientists were already packing their bags, and, for the most part, I could not help being pleased. But many more would have to go, and that would be painful. I suspected that granting agencies would be an ally in change, for it was clear that cell biology was "in" and classical chemistry, biochemistry, and physiology were, nationally, less favored for support.

The big question for me in 1970 was whether the Foundation could attract first-rate scientists. There was certainly no shortage of talent in the country. But while two decades of generous support had produced a new generation of young scientists eager to get to work, they were different in one respect from those of my generation: they were showing much more anxiety over funding shortages. This increased concern about personal security could make the Foundation less attractive than a well-endowed university, even if we made salary and research support more appealing. Traditionally, scientists had tended to favor major urban settings where other institutions and cultural attractions were readily at hand. There was evidence that this was changing, that scientists were seeking less complex and frenetic work environments. But our problem was to attract people unafraid of a *changing* institutional environment, as we moved away from more traditional areas of research to ones in which we had no proven institutional track record.

There was one additional disquieting personal thought. I had agreed to take the directorship primarily as a matter of conscience—the institution was my father's creation, and I could not let it perish without trying to help. In a way, I envied the several candidates who had turned the job down. They at least had a chance to make the decision objectively, if such decisions are ever made objectively.

While it serves my sense of drama to emphasize the odds against me as I began my job at the Foundation, there were in fact some very solid assets. First, right from the start and throughout my fifteen years, the Foundation's board of trustees was an unshakable bulwark. Many of them had been brought up in an enlightened tradition of philanthropy with respect to Worcester's cultural institutions. Many, too, had become good friends of my father and Pincus, both of whom had from the beginning been a part of the life of the community. The trustees came to cherish the two scientists not only as companions but also as benefactors of society. The feeling of goodwill and satisfaction in the success of a worthy enterprise that my father and Pincus had woven into the fabric of the institution seemed to have been generously transferred to me.

Another asset was the structure—or rather, the lack of structure—of the institution. It encouraged change. The absence of departments, the single-minded commitment to research, with teaching limited to the supervision of postdoctoral fellows, and the almost complete dependence on grant money as opposed to endowment income encouraged institutional flexibility and responsiveness to intellectual initiatives for change. I suspect, too, that these attributes enhanced the institution's appeal to a young faculty for whom *venturesomeness* and *unorthodoxy* were not bad words. For me, certainly, the word *director* had real meaning: I was at the controls, not enmeshed in a straitlaced bureaucracy, as a president or dean of a university would be.

A third favorable circumstance was the largely voluntary departure of old-guard scientists before, during, and after my arrival. Although their loss contributed to financial problems because of the loss of indirect cost allowances on their grants, and to morale problems among the remaining faculty, it did make space and facilities available for scientists in new areas of research.

Federico Welsch, ca. 1980 *(Worcester Foundation archives)*

Finally, the institution was blessed with a superb administrative staff already in place: Bradford Bull, financial manager; Richard Eastman, business manager; and Paul Kennedy, purchasing manager. These people were not only highly competent at their jobs but a pleasure to work with, and they became my good friends. Later on, we were to add a fine development officer, Robert Hyde, who elevated fund-raising to a respected professional activity and taught many of us the intricacies of the art. In addition, a colleague of special value, Federico Welsch, came to the foundation with me as executive director. Federico had been a graduate student in biochemistry at Dartmouth Medical School, having already obtained his M.D. at the University of Valencia in Spain, and had served on medical school faculties in Mexico before coming to Dartmouth. He proved to be a valuable associate in my efforts to reorganize the teaching of biochemistry; when I decided to accept the Worcester Foundation job, I asked him to join me. His managerial and fiscal talents, monumental capacity for work, and enthusiasm substantially eased my administrative burden. It speaks for the harmony of our "family" that these men and many more in administration stayed on throughout my tenure, in spite of a 70 percent turnover in faculty.

In a dramatic demonstration of the effectiveness of the new management team, Federico and Brad Bull tackled the Foundation's growing operating deficit, which had reached $250,000 by 1970, by cutting operating costs and improving recovery of indirect costs. They balanced the books within one year.

The year 1971 marked the inauguration of President Nixon's war on cancer and a substantial increase in the National Cancer Institute's (NCI) budget for research. As part of the new cancer initiative, which in the coming years was to reach a commitment of over one billion dollars per year, the NCI was establishing new cancer centers around the country. While most of these were to be *comprehensive* centers that combined research, teaching, and patient care in medical school–hospital complexes, so-called *specialized* centers were being created in universities and research institutes to promote basic research on the nature of cell growth and cancer. I believed our best chance for setting the Foundation on a new course and for bringing in substantial new support was to take advantage of these developments. The designation of the Worcester Foundation as a cancer center would mean core support for the salaries of key people, equipment, and facilities. It would also mean the kind of prestige that would attract additional support from private donors.

My research "homeland," the Huntington laboratories at the Massachusetts General Hospital, had been a focus of much of Harvard's cancer research program, under the guidance first of Joe Aub and later of Paul Zamecnik. As I have noted, these men had broadly interpreted cancer research as an effort to understand both normal and abnormal growth and development, expressed since the late 1940s by the focus on endocrine function in relation to cancer and the work on protein synthesis. My Huntington background had thus identified me with an enlightened, basic approach to cancer research.

I was indebted to the Huntington, both because Aub and Zamecnik's work had had an important impact on national cancer research trends and federal policy and because it supplied a vital institutional model. Cell biology had a special relevance to the problem of cancer, all the essential manifestations of which could be seen as properties of single cells. The genetics of cancer could thus be explored in cultured cell populations, as could the actions of such causative agents as viruses and chemicals. My own interest in mechanisms of growth and regeneration and some aspects of the Foundation scientists' interest in reproduction and endocrine regulation were certainly relevant. Cancer was ripe for the molecular attack. As a pure scientist wishing to rebuild a pure-science institution, I felt fully justified in linking up with the national cancer effort.

Over the years at Harvard, I had had congenial contact with Sidney Farber, who had done much to mold federal cancer policy. Farber was also a friend of my father and a trustee and booster of the Worcester Foundation, so I sought Farber's opinion about our chances of getting the Foundation designated and funded as a specialized cancer center. Farber surprised me by his enthusiasm, which seemed disproportionate to the modest beginning I had to offer. He encouraged me to make the attempt. He was sympathetic in part because he had for years fought departmental parochialism and obstructionism at Harvard Medical School in his own attempt to install an interdisciplinary cancer research program there. He finally succeeded, but only outside Harvard's formal academic framework, at the Children's Hospital, where his efforts created the Dana-Farber Cancer Center. I suspect he was intrigued by what a truly open institution might be able to do in establishing a coordinated research program in cancer research. I was emboldened by his faith and saddened that his death soon afterward prevented his enjoying our success.

In order to qualify for cancer center support, a significant nucleus of qualified people had to formulate a coherent, novel plan of attack on a set of crucial biological problems relating to cell growth. There were no incumbent scientists whom I could imagine in a new cell biology program. During the Dartmouth interlude, however, I had negotiated with a biophysicist, Donald F. H. Wallach, and our mutual interest in fashioning new approaches to cell growth drew us together again. He agreed to become a part of the cancer center effort at Worcester and to help us put together a strong proposal. His interest in cell surfaces, in the molecular characterization of cell membranes, in membrane assembly, and in hormone receptors on cell membranes was highly relevant to cell biology and cancer and complemented my interest in macromolecular synthesis and its regulation. We envisioned a program in which fundamental problems of cell identity, cell-cell interaction, and cell growth would be integrated in the hands of a group of like-minded investigators. We plunged heartily into the writing of a major proposal to the NCI.

I also approached my two most able former postdoctoral associates, Oscar Scornik and Sam Wilson. Oscar, as I noted, had settled at Dartmouth. Sam had gone to the National Cancer Institute in Bethesda, Maryland, and was rapidly making a name for himself in the investigation of cellular mechanisms of DNA synthesis. Their areas of research were highly appropriate for the envisioned program. I asked each of them to consider joining the Worcester Foundation, at least with sufficient commitment to be willing to write research proposals to be integrated into a grand package. They agreed. Another scientist, Keichi Hosokawa, who was interested in the structure and function of ribosomes and who had earlier considered a post in the department at Dartmouth, also was willing to join our effort. The five of us constituted what we considered a nucleus of expertise in cell structure and macromolecular synthesis. We wove those interests together into a proposal that described an integrated, basic research approach to an understanding of cell growth control. In its introduction, I wrote,

> The philosophy behind the program envisioned has in part been generated by the extensive experience of [the participants] with medical schools as research environments. The rigidity and essential autonomy of

classical basic science departments make a true interdisciplinary approach to a single large biomedical research objective extremely difficult. The institute framework on the other hand, with a centralized basic research–oriented administration, permits the defining of goals and then setting about obtaining the best

Samuel Wilson, ca. 1986 *(courtesy Samuel Wilson)*

> possible expertise regardless of formal academic affiliation. There is no need to be concerned with a proper "balance" of disciplines, the prime concern being whether the disciplines concerned contribute to the elucidation of the problem in hand.

We understood at the outset that our joint effort was tentative and provisional. The NCI understood this as well. If support was forthcoming, each participant would have to decide whether to leave his respective institution to become a part of the new effort.

On a fateful day in the spring of 1971, a committee of scientists assembled by the NCI arrived at the Worcester Foundation for a two-day evaluation of our proposal, which the members had studied in advance. As they gathered in a Foundation conference room, we received a staggering blow. Wallach, the only scientist who had firmly contracted to join us and whose scientific interests had strongly influenced the character of our grand plan, arrived late, after all the other actors in the drama had assembled, took me aside, and with trembling hands and red face announced that he was abandoning the enterprise. He had in fact already accepted a job at another institution.*

Deeply embarrassed and humiliated, I had to announce to the committee the defection of a star performer. This produced the second surprise of the day: the committee

---

*It emerged that Wallach had misgivings about the administrative structure I proposed for the cancer center. He wanted the center to be a *separate* administrative unit within the Foundation, with its own director, which he aspired to become. I had wanted the activity to be, at least in its early stages, an integral part of the Foundation's total research effort, under my directorship. The latter arrangement was favored by NCI as well, and it turned out in practice to be highly satisfactory. In particular, it led to harmonious relations between the new cell biology faculty and the rest of the Foundation family. Near the end of my stay, in preparing for a transition in leadership, Thoru Pederson became the center's director, without, however, changing its relation to the institution as a whole.

had had full knowledge of the situation. The university Wallach was joining also was establishing a cancer center, and the NCI had informed the present committee that Wallach had already agreed to be a member of it. In an incredible display of goodwill and confidence, the committee assured me that our proposal's merits transcended actions of individual scientists. Its favorable impression of our program was conveyed to a parent centers committee and from there to the council of the NCI. We were funded and began operating as a specialized cancer center in the fall of 1972. We were the first of seventeen such centers to be established under the new cancer initiative.

Only one of the original group of scientists named in that initial application, Hosokawa, has ever actually become a part of the cell biology effort at the Foundation. This was hardly unexpected in view of the tentativeness of these first steps in developing our cell biology program. One exceptionally able young associate of Wallach's, Grant Fairbanks, threw in his lot with us. He added to our strength in membrane biophysics and has since remained a strong contributor to the center's overall program. Sam Wilson remained at NIH and Oscar Scornik at Dartmouth. I am profoundly grateful to both of them for having given critical help in the initial effort.

Support for the center has been regularly renewed and expanded since 1972, and it continues to provide essential core support for the Foundation's cell biology program. Upon each renewal, we had to defend our stance before some critics who wished our research were more directly applicable to cancer cells. I take a special pride in having built a sound program from such modest beginnings and also am gratified that our strong commitment to basic research continues to be upheld by the NCI.

While we eagerly took advantage of new federal sources of support for research, many of my colleagues and I were at the time not very sympathetic to Nixon's cancer initiative, which set the NCI apart from the other NIH effort and provided it with a budget far exceeding those of other deserving institutes within the NIH. The national cancer plan indeed generated a lot of bureaucratic silliness. Elaborate plans, illustrated with intricate charts, were seriously promulgated by NCI officials who portrayed the conquest of cancer as an assortment of scientific disciplines arrayed in a circle, aiming at THE CURE in the center, like Indians attacking a wagon train. We expected that new money would flow into patient care in the nation's hospitals and not enough into the search for understanding cell function. Fortunately, the NCI heeded the advice of those who urged further support of basic cellular research not too specifically targeted to narrow objectives. Cancer was—and remains—a profound mystery, the understanding of which required exploration of uncharted domains by investigators free to pursue the best leads, wherever they were clever enough to uncover them. As the NCI's annual budget rose to over a billion dollars during the 1970s, the proportion of funds designated for basic research remained relatively stable.

A year or more into the job, with the deficit eliminated, with reliable and imaginative administrative help, with the support of NCI and the first members of a promising faculty on board, and with the generosity of the central Massachusetts community, I was enormously emboldened. I had a sensation of decompression, of a release from the heavy constraints of academia. The Foundation reawakened my warm memories of the Huntington, the Carlsberg, the MRC group—memories that had slumbered during the medical school years at Harvard and Dartmouth. I had an exhilarating sense of freedom and of being able to make things happen. For the first time, I really understood my father's and Pincus's euphoria in abandoning academia to start a new experiment in the nurturing of basic science.

9

✣ ✣ ✣

Rejuvenation

In looking back over my career, I can say that while I sometimes lacked confidence in my durability as a practicing scientist, I did have faith in my ability to *judge* good science and good scientists. I was prepared to believe that it would take time to attract first-rate scientists to the Worcester Foundation—I was even prepared to believe we never

Thoru Pederson, 1985 *(courtesy Thoru Pederson)*

could—but I knew I would at least recognize them. Among our early recruits, to my lasting satisfaction, was Thoru Pederson, who later became my successor. His excellent training in molecular modes of gene expression, clearly at the front of advancing research, ideally fitted him for a key role in the cell biology effort. He wanted to do full-time research and had a beguiling spirit of adventure, unconventionality, and an indifference to career security. His brilliance and originality, his high standards in the judgment of people and projects in science, and his conviction that the Foundation was going to be on the map in molecular and cell biology made him a superb ambassador. He raised my spirits as nothing else could. His appointment made me realize that the Foundation would appeal to a special subgroup of excellent scientists—namely, those who truly wanted to devote themselves full-time to research, who sought a simpler setting, and who, in general, had sufficient confidence in their potential to be less seduceable by the panoply of academia. Pederson quickly set up a lively research program, attracted able postdoctoral students, and entered into the institutional planning process.

We knew we would survive and thrive only by attracting outstanding scientists who would be able to bring with them, or soon generate, much of their own support. We also knew that, at least at first, we would have to build with young scientists, because established scientists would be reluctant to forsake university tenure and move to a less secure environment. This meant having intellectually challenging programs, an appealing setting, a growing nucleus of bright and enthusiastic scientists, and reasonable financial stability. The latter required a steady and dependable generating of some 10 to 15 percent of the Foundation's total income (mostly grants to individual scientists) from private sources. In other words, we had to double our annual private income to about one million dollars.

To supplement our efforts in connection with the cancer center, we launched a major private fund-raising campaign specifically in support of our new efforts in cell biology. It was gratifyingly successful. The money raised enabled us to recruit faculty and guarantee their salaries and research support for one or two years, until they could obtain their own grant funding.

A youthful faculty would mean a relative shortage of older mentors to encourage and guide younger scientists. This would have to be compensated for by the institution's promotion of wide communication between scientists, by the encouragement of visiting-scientist and seminar programs, and by an intensive collaboration with scientists at other institutions.

We believed it entirely appropriate to continue to strengthen the Foundation's programs in reproductive and endocrine biology and in neurobiology, as long as the orientation moved toward cell biology. This entailed, as I have said, a considerable turnover of personnel. We had to look at ourselves as a relatively high-turnover institution anyway—why not make the most of it? If we could keep our flexibility and responsiveness, *and* attract first-rate faculty with guarantees of reasonably stable salaries and generous support for research, we would have the best of both worlds. These decisions were shared with and approved by a nine-member external scientific advisory board that helped us evaluate our planning from 1971 to 1975.*

---

*Members of this board were the following: David Baltimore, professor of microbiology, Massachusetts Institute of Technology; Konrad E. Bloch, professor of chemistry, Harvard University; Edward A. Boyce, member, Sloan-Kettering Institute for Cancer Research; Nancy L. R. Bucher, associate professor of medicine, Harvard Medical School; Oscar M. Hechter, professor of physiology, Northwestern University; David H. Hubel, professor of neurobiology, Harvard Medical School; Elwood V. Jensen, director, Ben May Laboratory for Cancer Research, University of Chicago; David Shemin, professor of chemistry, Northwestern University; and Paul C. Zamecnik, professor of oncologic medicine, Harvard Medical School.

There prevailed among incumbent faculty at the Foundation an attitude that the function of the administration was essentially custodial: it should not set research directions and goals but simply ensure the well-being of a faculty that had already established its territorial claim by virtue of having obtained grant support. And in recruiting new faculty the administration should only augment existing research programs. Our national research support system, which directly supports working scientists after a review of their research plans by committees of peers, encourages this view. And it had been passively nurtured by the Foundation's previous leaders in the rich soil of expanding federal support for basic science during the 1950s and early 1960s.

No scientist with his own grant support had ever been asked to leave the Foundation. There were no formal institutional mechanisms by which a scientist's value to the institution could be measured. There were no formal contracts between the institution and its faculty with regard to salary, research support, or space and no formal job descriptions. Titles had been awarded rather arbitrarily and often were meaningless, and perquisites were poorly delineated. Nor were there rules that obliged scientists to communicate their ongoing work and research plans to other scientists within the institution. The result was a loose community of independent investigators without formal guarantee of institutional support, sometimes working at cross-purposes and instinctively erecting the kinds of barriers to communication that academic institutions already had formalized—owing largely to a lack of conscious administrative effort to prevent such developments. And, as the rising curve of government funding leveled off in the few years before I came, faculty fears of being left high and dry, without significant institutional backing, rose sharply.

One of the first changes we made was simply to require

that all grant applications from the Foundation be reviewed by the director and at least two other scientists before submission to a granting agency. Almost overnight, a greater openness among the faculty was created. We introduced a university-like appointment system with three-, five-, and seven-year-term guarantees of salary, made with the expectation that new appointees who did not have their own grants would get them within one or two years. We also established three university-style tenure appointments, immediately awarding them to the most deserving incumbent faculty members. The purpose was to boost morale and, since two of the recipients were not too far from retirement, to have slots available in the not too remote future. We assumed that we would be building with a young faculty, so open tenure posts would not be essential in the early phases. Later, we increased such lifetime tenure commitments to six.

In many universities, more than two-thirds of the faculty is on lifetime tenure, a state of affairs that undermines, it seems to me, the intent of tenure and helps entrench mediocrity. If the Foundation had enjoyed an adequate endowment, we might have taken the same route, but I like to believe that we would have had enough wisdom to limit the number of such appointments. While I opened my mail every morning for fifteen years praying that someone had made a gargantuan bequest to our endowment, I think our relative poverty was good for us.

As a result of our determination to change research direction, new rules of appointment, and a vigorous recruiting effort, only ten of the thirty-six scientists on the staff in 1970 remained in 1985. Of those, three, as I noted above, had been awarded tenure, two opted for and received term assignments, and five chose to remain with their pre-1970 status. Of the twenty-six who departed during my fifteen-year stay, none opted for any term appointment.* We were enormously proud of this dramatic turnover since it meant the achievement of our cell biology objectives and was accompanied by a steadily rising spirit of institutional optimism.

By 1985, there were sixteen scientists in our new cell biology program. The group realized our hope of establishing an integrated program in the investigation of gene expression; the organization and function of cell structural elements, particularly microtubules and membranes; viral carcinogenesis; and molecular mechanisms of cell development. With funds raised from the central Massachusetts community and other private donors, and an additional grant from the NCI, we built a new research building to house the group.

The cell biology group came to constitute about 40 percent of the Foundation's total personnel and a proportionate share of its $10 million annual budget. The harmony of the group contributed to and reflected the program's true interdisciplinary character. The group also came to be, as we had hoped, an effective stimulus in altering the perspectives of the Foundation's other two research groups.

There were twenty-three scientists in endocrine and reproductive biology in 1970. By 1985, their number had dropped to fifteen. Five of these were of the original group, and ten were new: five of them bona fide cell biologists and five representing a largely unsatisfactory compromise between our new institutional objectives and the scientific

---

* Like new faculty, incumbent members were offered the option to undergo review for term appointments or to remain with their pre-1970 status: self-supporting on grants with a one-year terminal guarantee of salary.

tastes of the incumbent faculty. An additional ten scientists had been recruited during the period but had left before 1985. Heirs of the Foundation's fame in reproduction research, the members of this group seemed to suffer a profound malaise at first. The new thrust in cell biology was a cause of listless resignation expressed in an ineffectual effort at recruiting. Part of the problem was the absence of a leader identified with *their* interests. While the thrust in cell biology strengthened their hand in terms of wider recognition of the Foundation's "rebirth," with its salutary impact on potential candidates and the granting agencies, this could not be a substitute for lost leadership.

In neurobiology, there were seven scientists in 1970 and nine in 1985. Six of those nine were recruited during my tenure, and all of those six were compatible with the objectives of our cell-molecular thrust. Again, an additional four scientists had been recruited during the period but had left before 1985. Thus, during those fifteen years, the research perspective of neurobiology shifted from an emphasis on the physiology of animal behavior and brain biochemistry in normal and pathological states to a more cellular, molecular, and genetic orientation. The group's focus is now on the analysis of electrical and chemical events in communication between nerve cells, the molecular basis of learning and memory, the effects of drugs at the molecular level, and the study of the genes and proteins that ultimately control events in the nervous system.

The readiness with which the neurobiology group responded to the initiative in cell biology was in part attributable to the relatively greater goodwill among the incumbent group engendered by my father, their patron. It was also directly influenced by Foteos Macrides, a gifted young neurobiologist who came to us from MIT in 1972. Like Pederson, he not only contributed by doing good science but with exceptional will and ability involved himself in institutional policy and planning and outside ambassadorial functions.

The process of restaffing the Foundation generally went far more smoothly than I had anticipated. I believe there were several reasons for our success. First was the openness of the recruiting process.* Second, the national system of science support favored the change in direction of research that we favored, and all our scientists knew it. Third, most of those who were unhappy with the new developments departed voluntarily.†

An important aspect of mobilizing initiative and willingness to change was the autocatalytic nature of the recruiting process: good scientists recognized each other. Thoru Pederson, Foteos Macrides, and others were immensely valuable in getting us off on the right footing. With the fresh air of the outside scientific world about them, with their relative youth and belief in the institution, and with their infectious enthusiasm, they painted a vivid picture of the institution to prospective faculty.

The policy of openness in recruiting began to pay off for the endocrine-reproductive biology group in 1980–1981

---

* Recruitment was the responsibility of three small committees representing the areas of our research interest. Each group was asked to keep the members of the other two groups informed of its efforts and to see that dossiers and candidates were seen by all concerned, when feasible. Final decisions on appointments were made by a consensus of the directors and a standing committee consisting of six elected scientists (subject to approval by the trustees).

† There was one distressing exception, a newly appointed scientist who brought suit against the Foundation on grounds of sex discrimination. The resulting fracas divided the Foundation into two camps during three critical years, 1972–75. It all climaxed in two court proceedings, which the Foundation won decisively. The scientist departed, and peace was restored almost overnight. We were deeply grateful for the astute, gentle, and perceptive leadership of Robert Cushman, president of the Norton Company and chairman of our board of trustees at the time, which helped immeasurably to ameliorate the situation.

when three appointments were made with the participation of the cell biology group. Lethargy vanished almost overnight. David Wolf, from Johns Hopkins University, was a physical chemist interested in the dynamics of membrane changes in cells, particularly in spermatozoa. George Witman, from Princeton, was an expert in the motility of cells, including sperm, and Charles Glabe, from the University of California, was studying sperm surfaces in relation to the fertilization process. We awoke to the realization that we had, in these scientists and others already recruited in cell biology, half a dozen scientists who could apply cell biology to the study of male reproduction, a traditionally neglected area. They, in turn, were surrounded by a growing cadre of scientists with closely allied interests in many aspects of cell structure and function.

In the early 1950s, M. C. Chang had discovered the phenomenon he named *capacitation*, the process by which sperm, initially incapable of fertilizing the egg, are made capable—are capacitated—by factors in the female reproductive tract. This remarkable phenomenon had received little attention since the emergence of molecular biology. Several of our people were now motivated to explore the process with tools and concepts never before applied to it. Chang's pioneering contributions to the field of in vitro fertilization also provided a lead-in to more molecularly oriented investigations of that process.

At my suggestion, the relevant scientists combined forces and prepared a major proposal for core support of a new, coherent program for the study of male reproduction. The proposal encompassed spermatogenesis, the various stages of sperm maturation in the male reproductive tract, sperm motility, capacitation, and fertilization. We assembled an outside committee of experts in reproductive biology that visited the Foundation and evaluated the proposed program; with its encouragement, we submitted a major grant proposal to the Andrew W. Mellon Foundation. We were successful in getting the support, and the Worcester Foundation's male fertility program has continued to flourish with generous support from the Mellon Foundation.

Another bright event in 1980 for the Foundation was our success in recruiting Paul Zamecnik. Paul had reached retirement age at Harvard, and the Huntington laboratories were to be replaced by a new Department of Genetics. After his feats in protein synthesis, he had continued to maintain a highly productive research program, exploring details of the interaction between amino acid–activating enzymes and tRNA, discovering a new compound of possible importance in the control of cell division and then moving into virology, where he was contributing to the development of antiviral agents. Notably, he opened the door to a new field—the investigation of certain short lengths

Paul Zamecnik and author *(left)*, ca. 1984

of DNA ("antisense" oligonucleotides) that inhibit viral replication. The use of these synthetic oligonucleotides in the treatment of AIDS is now under intense investigation in Paul's and many other laboratories. His vast knowledge of the molecular landscape and his wisdom and generosity in helping younger associates have greatly enriched the Foundation.

As our cell biologists shaped our new programs, I became aware that I was not going to be on the research team as I had planned. From the time I moved to Harvard Medical School in 1960, I had published only eighteen original papers, representing mostly the imaginative input and hard work of my younger associates. Over that period, I had a growing sense that my contributions were not up to my own standards of originality. After I came to the Worcester Foundation, five more papers were published, representing work that had been done mostly at Dartmouth. Only one new paper arose from work done at the Worcester

Hudson Hoagland and author *(left)*, ca. 1985 *(Worcester Foundation archives)*

Foundation.[1] So in 1974 I closed my laboratory and concentrated on the job of running the Foundation. I came to see in my administrative role a unique opportunity to make the Worcester Foundation a model setting for the encouragement of good science. In addition, troubled by the nation's waning support of science, I wanted to argue the cause of basic research before a wider audience. With steadily increasing enjoyment, I started to write for the public: a biweekly newspaper column in the Worcester papers and two books on biological science.[2]

My father had been relieved at the resolution of the directorship problem and seemed pleased to have me there. He voiced his pleasure widely, and the goodwill his enthusiasm kindled was helpful to me in my task. He was so firmly identified with the institution in the community that I was often confused with him. Indeed, I suspect that some people never knew there were two Hoaglands at the Foundation and attribute to one man the accomplishments of both.

When I first came to the Foundation, I had received the usual gratuitous advice on how to deal with a retired chief executive who wanted to stay on, especially if he was one's father. Its tenor was that, since there was no "upstairs," he should be persuaded to move out. I encouraged him to stay at the Foundation, however, and to continue to occupy his office, and I have never regretted my decision to do so. During the next twelve years, he was companionable and supportive, relinquishing the reins of administration gracefully and enjoying the deference accorded him as patriarch by trustees and staff.

My mother had been ill for many years and died in 1973. My father's life after that was heavily shadowed by the loss. He and I drew closer together in those years before his death, in 1982—through our shared cultural values and our love for science and for the institution we both served. There remained traces of awkward, self-conscious competitiveness between us—as though the parent-child duel must persist to the end. I learned from others how proud he was of me, for it was hard for him to tell me directly, but he clearly took pleasure in seeing a new generation of scientists thrive in the institution he had founded.

The Worcester Foundation has contributed, I believe, far out of proportion to its small size, to the advance of basic knowledge of life processes. Before 1970, major advances were made there in the elucidation of the secretion dynamics, metabolism, and mode of synthesis of cholesterol, and also of the adrenal and sex steroid hormones that are derived in the body from cholesterol. The development by Foundation scientists of methods for transplanting ovaries and uteri in sheep, permitting a more controlled study of endocrine dynamics, has enhanced our understanding of reproductive processes, including the role of the prostaglandins, a whole new class of hormone-like substances that was discovered in the 1960s. The contributions of M. C. Chang and his associates to the understanding of the fertilization process in mammals, including the first successful in vitro fertilization of rabbit eggs, and the discovery of capacitation have been widely hailed. And Chang and Pincus's discovery of the ovulation-inhibiting properties of the progestins is a monumental achievement in biomedical history.

Since 1970, basic research on cellular and molecular aspects of nerve and brain function, on nervous system–hormone interactions, and on hormone receptors on cell surfaces and within cells has become increasingly probing. Major new insights have been gained into the male reproductive process. Foundation scientists have contributed in important ways to understanding the mechanisms by which viruses induce cancer, and they have been in the forefront of research on gene structure and expression in mammalian cells. Finally, they have contributed to the molecular characterization of cell membranes and cell organelles and to the understanding of the process of cell division.

In the mid-1970s, in the early days of recombinant DNA technology, Foundation scientists were actively involved in its use. Anticipating adverse public reaction to alarmist-generated scenarios featuring the production of dangerous microorganisms, we worked closely with elected officials and citizens in the Shrewsbury-Worcester community to inform them of our research and our safety precautions. I believe this educational effort was largely responsible for the relative equanimity with which the introduction of the new technology was accepted.

In addition to the achievements of its scientists, the Foundation has drawn attention to the value of flexibility, openness, and collegiality in a setting where creative people can devote essentially all of their time to the search for knowledge.

Biomedical science is shadowed today by a growing public suspicion that the search for knowledge is tainted by the ill uses to which it might be put by society. Impatience with slow progress against disease, blind opposition from animal rights and right-to-life groups, fear of "genetic engineering," and a pervasive and profound ignorance of the ways of science all contribute to the current atmosphere of mistrust. Scientists' own timidity and their failure to attack flabby government planning and their growing infatuation with product-oriented, profit-motivated research exacerbate the problem. In this atmosphere, we at the Worcester Foundation took great pride in being able to reawaken the Hoagland-Pincus vision of independence—and to reaffirm our own—in a spirited commitment to free exploration for its own sake. We shook ourselves free of an earlier passive acceptance of contract research and commercial influence. We were able to put the spotlight on young investigators who tackled fundamental problems. And we did it all without an endowment and during a time of waning federal support for basic biomedical science. There was a dimension of grace, of sharing, and of generosity at the Foundation that reminds me of the Huntington under Joseph Aub and Paul Zamecnik. We came to feel that this kind of institutional environment is as close as science is likely to get to the ideal setting for exploration.

10

And Some Politics

The practice of science is a prerequisite, but not a preparation, for its administration. Science deals with things; administration, with people. The move from one to the other is often guided by the principle that if you do a job well, you must "rise" to a job for which you have no demonstrated competence. This opportunity comes at the right time for many of us: in late middle age, as troublesome doubts about our creative ability snap at our heels.

I found the transition relatively easier than I had expected. It was a relief to be free of the close, objective scrutiny of my formidably competent colleagues. My productivity would henceforth depend more on my skill in judging broad scientific trends and the qualities of younger colleagues and on personal interactions and politics. Administration also appealed to my predilection for theatricality, to which I have already alluded. It offered me an opportunity to perform a new part in the drama of science. I perceived the "saving" of the Worcester Foundation in that light. I passionately believe in the social value of scientific inquiry for its own sake, and I was being offered a new stage from which to shout my conviction.

The experience had negative aspects, of course. Administration is subtly insulating and isolating. Although scientists appear to be loners, the hours they spend generating and nursing ideas, pondering problems, assessing strategies, and doing experiments are shared with like-minded colleagues to an extent not generally appreciated by the public. Science fosters wide-ranging and profound friendships because of a fraternal preoccupation with issues that often cannot be solved in intellectual isolation.

The move into management brings with it a new set of companions who have not shared the experience of exploration and who speak a very different language. Moreover, as a boss, you experience subtle changes in your relationship with those around you. The usual hostility to authority remains largely hidden. Most of your associates seem to become increasingly agreeable; they say yes much more often. Conditioned to skepticism and agrumentativeness in your colleagues, you may be pleased at this evidence that you have apparently become more consistently right about all sorts of things. But these early signs of the corrupting effects of power bring to the scientist turned administrator, in his more candid moments, a sense of lost innocence.

I learned some useful things from administration that my sheltered life as a scientist had not required me to make much use of: the need for tolerance and for compromise, ways to mask boredom and to simplify, and especially the value of procrastination. I had been raised to deplore putting off until tomorrow what could be done today. And science, as action, finds little virtue in temporizing. But in dealing with people the quality of one's opinions and decisions can frequently be improved by delay, by playing for time, by simmering one's immediate responses in the juices of reason.

Of course, the big attraction that outweighs all the negatives is the opportunity administration affords for unique achievement—especially in an institution whose policies and structure have not become fossilized by convention. As I have said, as an ordinary scientist you may get *priority* for a discovery, but it is hardly unique. If I had not been there, someone else would have discovered what I did—not within years but within months. In the management of an institution, on the other hand, as in art, achievement is identified more closely with the performer.

There was one sense in which I *was* prepared for the job. I had already logged twelve years in research institutes. The Huntington, while formally part of Harvard's Department of Medicine, was in reality a research institute. Its staff pursued science full-time, did little or no formal teaching (except of postdocs), and had mostly non-tenure-track university appointments. The same was true of Lipmann's labs at the MGH. The Carlsberg Laboratory in Copenhagen, the MRC Unit in Cambridge, and the Pasteur Institute in Paris also were pure research institutes. All of these places, like the Worcester Foundation, were wide open and single-mindedly committed to exploration. The nine years I had spent in formal university departments had been rewarding in terms of contact with students, but otherwise constricting to the spirit. The dynamic of the research institute—its responsiveness, flexibility, and openness—mirrored the dynamic of science itself.

Most of the money for the support of basic biomedical research in universities and research institutes in the United States comes from the federal government via the National Institutes of Health. Since the mid-1960s, the budget of that enlightened and remarkably efficient agency has not kept pace with the nation's exceptional capability in basic research. A succession of presidents, spurning advice in regard to science, have fallen into the habit of offering budgets substantially below that required to sustain a stable, vigorous enterprise, knowing that Congress will restore most of the cuts. Insufficient funding is bad enough but can be tolerated. More frustrating for a community of committed long-term explorers is the unpredictability and instability of support. Such short-sighted and unplanned management of a vital national enterprise, in which the administration's powerful and scientifically ignorant Office of Management and Budget too often calls the shots, has become increasingly exasperating to the scientific community and, indeed, to many members of Congress.

During the early 1960s, Sidney Farber, along with the influential surgeon Michael DeBakey and the activist-philanthropist Mary Lasker, worked to good effect favorably to influence legislation and appropriations on behalf of medical research. They interacted productively with such health-oriented lawmakers as Congressman John E. Fogarty and Senator Lister Hill, who had taken leadership roles in matters of medical science. But during the late 1960s and the 1970s science-government relations had steadily deteriorated, and as a result support for biomedical research had languished.

My mounting concern over this state of affairs needed expression. I was reluctant to plunge into the political arena—like most of my colleagues, I was profoundly skeptical that scientists could influence politicians. But I was halfway there as head of an institution almost wholly dependent on federal funds.

And Olley pushed me hard, playing skillfully on my conscience. Didn't scientists have an obligation to get into the fray rather than sit by and watch free inquiry go down the tube? Wasn't fighting for basic biomedical research in itself a powerful personal antidote to one's growing cynicism about the political process?

In the fall of 1977, I went to Washington with three colleagues to explore the extent to which Congress might be interested in consulting with the biomedical science community. The idea behind our visit had germinated at MGH's neighbor institution, the Eye Research Institute. Richard Pharo, the administrative director of ERI, and Bradie Metheny, its public relations consultant, disturbed by the faltering national initiative, aware of Congress's traditional sympathy for health research, and exasperated by the scientific community's apathy, had determined to take action. Dr. Charles Schepens, director of the ERI, Dr. Fred Stone, a vice-president of Boston University and former director of the National Institute of General Medical Sciences, Bradie Metheny, and I made up the visiting group.

Bradie is a robust, affable, optimistic Missourian. Direct, honest, and politically savvy, he is well liked and trusted by the members of Congress. He had built up extensive contacts with congressional members and staff through the work of his small consulting firm, Tricom, and was ideally suited to coordinate our effort. I was invited because I was president of the Association of Independent Research Institutes at the time, an organization representing some forty of the nation's independent biomedical research institutions. I had also written a pamphlet on the importance of basic biomedical research called *Of What Use Is a Baby?*, which had received a wide circulation among professionals, media people, and legislators, and my first general-audience book on basic biology, *The Roots of Life,* was about to be published.

During our initial, three-day visit, we talked with some forty senators and congressmen and their staffs. To our astonishment, they were attentive and questioned us rigorously. As we moved from one appointment to the next, we found that our views had preceded us by phone, resulting in a continuing lively dialogue. Never before, we were told, had a group of professionals taken the time on their own initiative to discuss broad issues concerning the impact

Bradie Metheny, 1988 *(courtesy Bradie Metheny)*

of basic research on medical progress. In the past, Congress had been subjected to hordes of professional and lay witnesses pleading support for research in special areas and for particular diseases—the disease-of-the-month approach.

We sensed we were on fertile soil. Enormously heartened by the experience, we offered to return in force for more extensive discussions and were assured we would be welcome. I undertook the task of assembling a cast for the next production. That charter group* was an interesting mix of scientific and medical notables who could address with clarity the favorable impact of basic research on health care practice and the need for continued vigorous support. One of the most gratifying experiences of my life was the alacrity and enthusiasm with which these men responded to the call for help, often in spite of a deep skepticism. It was as though the conscience of the biomedical science community in general had been poised for action. Once enlisted, they worked with dedication on a completely voluntary basis. We named ourselves the Delegation for Basic Biomedical Research.

For the first year, Bradie put in long hours with consummate skill and little pay. Federico Welsch, with his exceptional talent as a fiscal manager, took on the staff role as "financial analyst." The Worcester Foundation picked up most of the cost of operation for the first couple of years. Later, annual contributions were obtained from most of the

---

*W. Maxwell Cowan, M.D., director, Division of Biological and Biomedical Sciences, Washington University School of Medicine, St. Louis; George R. Dunlop, M.D., University of Massachusetts Medical School, past president, American College of Surgeons; Seymour S. Kety, M.D., professor of psychiatry, Harvard Medical School, Boston; Arthur Kornberg, M.D., professor of biochemistry, Stanford University Medical School; Francis D. Moore, M.D., professor of surgery, Harvard Medical School; George E. Palade, M.D., professor of cell biology, Yale University School of Medicine; Charles L. Schepens, M.D., Eye Research Institute of Retina Foundation, Boston; Lewis Thomas, M.D., president, Memorial Sloan-Kettering Cancer Center, New York; James D. Watson, Ph.D., director, Cold Spring Harbor Laboratory, New York.

nation's independent research institutes and from medical schools and hospitals where we had contacts. By the mid-1980s, some forty institutions and foundations were contributing over $100,000 annually to our work. The monies were placed in a tax-exempt fund from which the delegation drew its expenses for staff and secretarial support. We operated, in effect, as an advising and consulting agency responsive to invitations from Congress or the administration.

The pattern of our work took the form of testimony before congressional committees, as well as separate meetings with key legislators and with administration officials. We were also called upon to provide written responses to questions by individual legislators. Congress's requests for information or advice could be counted upon to be at short notice. With Bradie's and Federico's input on political and fiscal matters, I would prepare white papers for hearings and written answers to queries and circulate them to our members with a deadline for a quick response. It was understood that my response, as spokesman, was to stand as an expression of the group's views if no amendments were forthcoming in the period specified. (This did not preclude members' supplementing our formal response.) The small size of our group, its prestige, its commitment to the *general* support of basic research without favoring any particular NIH institute or disease, and the rapidity of our response to queries or requests to appear for hearings made us a particularly effective advisory group.

Our constantly reiterated theme was the need for a consistent, forward-looking national health science policy, with annual predictability in the level of funding. We stressed the need to reaffirm the value of the principal vehicle for science support—the peer-reviewed, investigator-initiated project grant. From the mid-1960s to 1978, when we began our work, the number of new project grants funded fluctuated wildly—sometimes by 100 percent from one year to the next—making it extremely difficult for the scientific community to develop a long-range agenda for exploration or to plan for the training of manpower. We therefore urged that the government commit itself to fund a fixed number of these grants each year (we recommended five to six thousand). This stabilization would represent the essential beginnings of a science policy on which new investigators entering the system could count: the doorway would be about the same width each year. It was a policy that Donald Fredrickson, then the director of the National Institutes of Health, with whom we kept in close touch, was also urging at the time. Such a commitment required substantial additional sums of money for the NIH budget each year, on the order of several hundreds of millions of dollars. We were gratified to discover that first Congress and then the administration was willing to make this commitment.

We pressed our views upon representatives in policy papers replete with anecdotes illustrating how untargeted, pure research had led to practical benefits in medicine. We conducted seminars on biomedical research progress for congressional staff and the press. We had feared that grant-supported basic research—as opposed to disease-oriented research—would not be seen by congressmen as a marketable commodity among their constituents, but we were wrong. Fed up with the disease-of-the month approach, most of the legislators reaffirmed their confidence in the NIH's proven, internationally emulated record of success in supporting science by focusing support directly on individual scientists. They responded positively to our arguments that the pursuit of knowledge without concern for immediate application had achieved immense benefits for practical health care in the past; that applied research and technology become ever more costly and ineffective if not accompanied by a growing fund of basic knowledge; and that relative to our nation's total outlay for health care, our expenditures for new knowledge were grossly inadequate. We were pleased to find that Congress incorporated such views into the language of legislation and appropriations bills.

One key to our success was Representative Joseph Early, Democrat of Massachusetts, a member of the House Subcommittee on Health and Human Services. Joe, whose home was Worcester, knew and admired the Worcester Foundation. A round, warmhearted Irishman with a modest manner, unusual candor and spontaneity, and a cryptic political savvy, he strongly encouraged the delegation's efforts and urged his fellow committee members to seek its advice.

I believe that the delegation helped Congress reaffirm the fundamental value of the peer-review system, whereby grants are awarded to scientists on the basis of merit. Congress always faces heavy pressure from states and institutions to circumvent the process and allocate funds on political, economic, prestige, or other nonmerit bases. The system is periodically attacked in Congress and by some professionals on the grounds that it encourages mainstream, "safe" research as opposed to innovative, risky research, but we insisted that no better way has yet been found to ensure fairness and quality in allocating funds. And we reiterated our conviction that the problem of conservatism is not intrinsic to the peer-review process but a result of the anxiety and pinched vision that restricted budgets always bring in their wake. Innovation and creativity thrive when support for research is generous and when those aspiring to a life of adventure in science see long-term opportunity.

The legislators could hardly fail to be impressed with the fiscal efficiency of the NIH in fulfilling its mandate to improve human health. We called attention to studies showing, for example, that for every dollar NIH spent, thirteen dollars have been returned to the economy in improved health,[1] and that technology derived directly from NIH-supported basic research without thought of profit has led to industrial applications worth tens of billions of dollars annually.[2] Finally, we talked a lot about the importance of basic research to the economic health of our nation and to its competitive stance within the community of nations. We stressed the view that the growing niggardliness of U.S. budgets for nonmilitary science was economic suicide.

The delegation had a significant role in effecting substantial increases in funding for the NIH from 1979 to 1986, in encouraging a greater consistency in federal science support policies, and in improving the climate for those aspiring to a career in science. We made some beginnings in a new mode of science-government interaction at a time when it was at an all-time low. Without our voice in the late 1970s and early 1980s, the scientific community would have had little effective input into federal legislation and appropriations for biomedical research.

Some aspects of the delegation experience were very frustrating. The political system is not geared to the kind of foresight in planning that science requires. The protracted period of scientific training and the inevitable uncertainty of experimentation require more assurance of long-range stability than do most human endeavors. We scientists could readily project, five and ten years ahead, the need for science for new facilities, replacement of obsolete equipment, and training of personnel. But such assessments could not be fitted into the restrictive, one-year time frames of the federal appropriations process. Each year was a new ball game in which the same arguments and stopgap solutions were disinterred and rehashed. We were repeatedly called upon to frame—as dramatically, colorfully, and anecdotally as we could—health science issues and needs that were tiresomely obvious to us. Always lurking nearby was the nation's preoccupation with an obscene arms race, consuming hundreds of billions of dollars a year when the expenditure of a few hundred million would have made an enormous difference to the biomedical research effort.

There was, and is, a great deal more to say, much of it having to do with America's need to regain its spirit for the search. In that pragmatic forum, we spoke with conviction, but somewhat hesitantly and self-consciously, of our belief that the spiritual health of a nation is measured by its inclination toward enlightenment and intellectual adventure—goals entirely worthy in themselves. A national reluctance to support basic science was both a cause and a result of inadequate funding of research. As funds tightened, the expansive spirit shriveled and aspiring scientists were discouraged. However, in those long dialogues with Congress, we were more persuasive when we justified the nation's encouragement of science on the basis of the compelling evidence that the pure drive to know is the wellspring of limitless technological advances in the service of society's urgent interests, particularly the public health. Delegation member Lewis Thomas has written:

> The history of medicine has never been a particularly attractive subject . . . [because] it is so unrelievedly deplorable a story. For century after century, all the way into the remote millennia of its origins, medicine got along by sheer guesswork and the crudest sort of empiricism. . . . Virtually anything that could be thought up for the treatment of disease was tried out at one time or another, and, once tried, lasted decades or even centuries before being given up. It was, in retrospect, the most frivolous and irresponsi-

> ble kind of human experimentation, based on nothing but trial and error and usually resulting in precisely that sequence. . . . It is astounding that the profession survived so long, and got away with so much with so little outcry.

And then, about 150 years ago, a simple kind of science crept into medicine. Physicians began to be able to acknowledge that what they did to their patients did not always work and that, if they left their patients alone, they would get well on their own. The focus of medicine changed to supportive treatment, careful observation, and accurate diagnosis.[3]

In the last few decades of this century, science and medicine firmly joined hands, transforming the latter into an effective agency for the relief of human suffering. We have witnessed applications of new knowledge of normal and disease mechanisms: advances in surgery; a vast array of technological innovations for diagnosis and treatment; the advent of antibiotics; remarkable successes of drugs used in the treatment of mental illness; heartening advances in understanding cellular and molecular mechanisms of cancer and genetic disease and of the body's ability to defend itself against disease; and a great deal about cultural factors contributing to disease. Indeed, present knowledge of how to prevent major illnesses such as heart disease, stroke, and cancer far exceeds our will to use it. We need to be constantly reminded, too, that behind all these victories was a century of maturation of the sciences of microbiology, biochemistry, genetics, and molecular biology—developing the basic knowledge upon which these applications rested. Science, in short, turned the physician into an effective agent in the promotion of health.

It seems to me almost beyond belief that when I began to do research in 1948 evidence was just emerging that DNA was the material of genes, and that only twelve years later we knew the genetic code, the structure and mode of DNA replication, and many of the details of how its information was translated into protein. We knew also that at the molecular level all living creatures were composed of the same basic building units, used the same genetic language, and were governed by remarkably similar processes underlying their fundamental life activities. It is no wonder that scientists who participated in this epic drama of exploration and discovery have long since come to the unswerving conviction that the free exploration for truth, unencumbered by any need for rationalizations of relevancy, will inevitably go on producing the means to subdue the ills that still afflict us.

Traditionally, the public has appreciated and supported science. Lately, though, biomedical science watchers have been getting noisier, warier, more suspicious, and more fearful—preoccupied with what they consider threatening precedents and flagrant abuses. They act as though every advance in science were designed for some illicit, dangerous, or immoral purpose. New agents of proven value are banned if they cause cancer in a mouse or a beagle hound. The mere mention of the word *clone* triggers visions of *The Boys from Brazil.* We haggle over the ethics of helping infertile people have children. We panic if a human egg is fertilized outside the body. We are shocked if a scientist seeks to learn from the study of human embryos. We accuse scientists of "playing God," or at least Frankenstein, if they consider applying genetic techniques to urgent medical and agricultural problems. We continue to fight the ridiculous battle against those who would outlaw the use of animals in medical research. We are even creating professors, courses, and university departments of biological and medical ethics, rooted in part in terrifying hypothetical scenarios peopled with mad risk takers and leering human experimenters. We hear again the strident cries of those who opposed vaccination for smallpox, fluoridation of water, treatment of venereal disease. The public's current response to risk/benefit and ethical issues raised by science becomes ever more knee-jerk, exaggerated, and undiscriminating.

The population of our planet recently reached five billion. Some experts predict it will double in the next half century. The energy we expend flailing at conjectural risks and hypothetical breaches of ethics could better be put to use on heroic new efforts to educate our citizens and on research to further our understanding of the world we live in and the world within us. We need knowledge, and we need compassion to fight *real* menaces: starvation, disease, ignorance, despair, and our preoccupation with weapons.

Among the major reasons for the public's focus on fear are the increasing complexity, pervasiveness, and "publicness" of science on the one hand and our nation's abysmally inadequate education of its citizens on the other—which results in an ever-widening gulf between the producers of knowledge and its consumers. Ignorance of the purpose and process of science leads to the notion that we can escape from fear and uncertainty simply by stopping this unsettling business of asking questions of nature. It might, in fact, be possible to do this as a nation because the instruments science needs in order to ask the questions are ever more expensive and because the funds for them can be cut off.

But to stop science would be to commit suicide. We have no recourse but to encourage the search for answers and wisely to control the *uses* to which we put the knowledge we uncover. We often do that very well—witness the constructive response of scientists, laymen, and government to the advent to recombinant DNA technology. Early apprehension was voiced by concerned scientists and laymen and fanned by an assortment of uninformed alarmists and misguided moralists. At scientists' request, a permanent federal recombinant DNA advisory committee was established with a diverse membership. Its deliberations and the accompanying public debate, in which well-informed medical people, scientists, and laymen participated, fostered the habit of rationality, which few seem now to want to give up. We have been able to define widely acceptable applications of the technology in medicine and are moving to do so in agriculture and in other areas. In the process, the horror scenarios are being recognized as imaginary, impractical, or impossible as we get on with the serious business of applying our new discoveries toward the betterment of mankind.

We can and do creatively hammer out pragmatic, pluralistic, and workable solutions.

J. Robert Oppenheimer once said, in speaking about the gap between science and the public, "I do not see how the scientist can evoke the same understanding and grateful warmth from his fellows as the actor who gives them pleasure and insight, and reveals their own predicament to them, or the musician or dancer or writer or athlete, in whom they see their talents in greater perfection, and often their own limitations and error in larger perspective." True enough. But science is adventure, discovery, new horizons, insight into our world, a means of predicting the future, and enormous power to help others. That ought to be heady enough stuff for anyone.

References

Preface

1. Peter Medawar, "Hypothesis and Imagination," in *Pluto's Republic* (New York: Oxford University Press, 1982), 130.

2. François Jacob, *The Statue Within: An Autobiography* (New York: Basic Books, 1988), 237.

3. Jacob Bronowski, *The Common Sense of Science* (Cambridge: Harvard University Press, 1978), 104.

4. James D. Watson, *The Double Helix* (New York: Atheneum, 1968).

5. Gunther S. Stent, *Paradoxes of Progress* (San Francisco: W. H. Freeman, 1978), 90, and chap. 5.

6. Max Delbrück, "A Physicist's Renewed Look at Biology: Twenty Years Later" (Nobel lecture, 10 Dec. 1969), reprinted in *Science* 168 (1970): 1312–15.

7. Jacob Bronowski, *Science and Human Values* (New York: Harper & Row/Perennial Library, 1972), 95.

1: A New Research Institute

1. Hudson Hoagland, *The Road to Yesterday.* (privately printed, 1974); Kenneth S. Davis, "The Story of the Pill," *American Heritage*, Aug.–Sept. 1978, 80–91; Carl Djerassi, "The Making of the Pill," *Science 84*, Nov. 1984, 127–129; Loretta McLaughlin, *The Pill, John Rock, and the Church: The Biography of a Revolution* (Boston: Little, Brown, 1982); James Reed, *From Private Vice to Public Virtue: The Birth Control Movement and American Society since 1930* (New York: Basic Books, 1978); Paul Vaughan, *The Pill on Trial* (New York: Coward-McCann, 1970).

2. Hudson Hoagland, *Pacemakers in Relation to Aspects of Behavior* (New York: Macmillan, 1935).

3. Hudson Hoagland, *Road to Yesterday,* 79–80.

4. Albert Q. Maisel, in Annual Report of the Worcester Foundation for Experimental Biology, 1967.

5. Hudson Hoagland, *Road to Yesterday,* 82.

2: First Steps

1. Anne Roe, *The Making of a Scientist* (New York: Dodd, Mead, 1952); idem, "A Psychologist Examines Sixty-four Eminent Scientists," *Scientific American,* Nov. 1952, 21–25.

2. Paul C. Zamecnik, "Cancer Research: Joseph Charles Aub," in B. Castleman, D. C. Crockett, and S. B. Sutton, eds., *The Massachusetts General Hospital* (Boston: Little, Brown, 1983), 343–48.

3. Nancy L. R. Bucher, "Dr. Aub, Huntington Hospital, and Cancer Research," *Harvard Medical Alumni Bulletin,* Fall–Winter 1987, 46–51.

4. François Jacob, *The Statue Within: An Autobiography* (New York: Basic Books, 1988), 219.

3: Vistas

1. Robert S. Grier, M. B. Hood, and M. B. Hoagland, "Observations on the Effect of Beryllium on Alkaline Phosphatase," *Journal of Biological Chemistry* 180 (1949): 289–98.

2. M. B. Hoagland, R. S. Grier, and M. B. Hood, "Beryllium and Growth: I. Beryllium-Induced Osteogenic Sarcomata," *Cancer Research* 10 (1950): 224–25.

3. M. B. Hoagland, "Beryllium and Growth: II. The Effect of Beryllium on Plant Growth," *Archives of Biochemistry and Biophysics* 35 (1952): 249–58; idem, "Beryllium and Growth: III. The Effect of Beryllium on Plant Phosphatase," ibid., 259–67.

4. Harriet L. Hardy, *Challenging Man-made Disease* (New York: Praeger, 1983).

5. Paul C. Zamecnik, "The Microsome," *Scientific American,* March 1958, 118; idem, "Historical and Current Aspects of the Problem of Protein Synthesis," *Harvey Lectures* 54 (1960): 256; idem, "An Historical Account of Protein Synthesis, with Current Overtones—A Personalized View," *Cold Spring Harbor Symposia on Quantitative Biology* 34 (1969): 1–16.

4: Amino Acid Activation

1. *The Carlsberg Laboratory* (Copenhagen: Rhodos, 1976), 21.

2. Ibid., 18.

3. Fritz Libmann, *Wanderings of a Biochemist* (New York: John Wiley, 1971), 4.

4. Fritz Lipmann, "Metabolic Generation and Utilization of Phosphate Bond Energy," *Advances in Enzymology* 1 (1941): 99 (reprinted in Bobbs-Merrill, Reprint Series in the Life Sciences, no. B-184).

5. Ibid.

6. See Paul C. Zamecnik, "An Historical Account of Protein Synthesis, with Current Overtones—A Personalized View," *Cold Spring Harbor Symposia on Quantitative Biology* 34 (1969): 6.

7. Paul Berg, "Participation of Adenyl-Acetate in the Acetate-Activating System," *Journal of the American Chemical Society* 77 (1955): 3163. See also Lipmann, *Wanderings*, 46.

8. M. B. Hoagland and G. D. Novelli, "Biosynthesis of Coenzyme A from Phosphopantetheine and of Pantetheine from Pantothenate," *Journal of Biological Chemistry* 207 (1954): 767–73.

9. Werner K. Maas and G. D. Novelli, "Synthesis of Pantothenic Acid by Dephosphorylation of Adenosine Triphosphate," *Archives of Biochemistry and Biophysics* 43 (1953): 236–38.

10. See Lipmann, *Wanderings*, 46–47, for his view of these events.

11. M. B. Hoagland, E. B. Keller and P. C. Zamecnik, "Enzymatic Carboxyl Activation of Amino Acids," *Journal of Biological Chemistry* 218 (1956): 345–58.

12. M. B. Hoagland, "An Enzymic Mechanism for Amino Acid Activation in Animal Tissues," *Biochimica et Biophysica Acta* 16 (1955): 288–89 (reprinted in H. M. Kalckar, ed., *Biological Phosphorylations: Development of Concepts* [New York: Prentice-Hall, 1969]).

13. Hoagland, Keller, and Zamecnik, "Enzymatic Carboxyl Activation."

5: *Transfer RNA*

1. As reported by Horace Judson, in *The Eighth Day of Creation* (New York: Simon & Schuster, 1979), 481.

2. See, too, Zamecnik's recollection of this period, in "An Historical Account of Protein Synthesis, with Current Overtones—A Personalized View," *Cold Spring Harbor Symposia on Quantitative Biology* 34 (1969): 6.

3. M. B. Hoagland, P. C. Zamecnik, and M. L. Stephenson, "Intermediate Reactions in Protein Biosynthesis," *Biochimica et Biophysica Acta* 24 (1957): 215–16.

4. See Zamecnik, "Historical Account," 6.

5. For a discussion of the adaptor hypothesis in relation to our discoveries, see Judson, *Eighth Day*, 268–70, 287–95, 313–28, 340, 412, 472.

6. M. B. Hoagland, M. L. Stephenson, J. F. Scott, L. I. Hecht, and P. C. Zamecnik, "A Soluble Ribonucleic Acid Intermediate in Protein Synthesis," *Journal of Biological Chemistry* 231 (1958): 241–56 (reprinted in *Perspectives in Modern Biology: Selected Papers in Molecular Genetics* [New York: Academic Press, 1965] and in *Protein Synthesis: Selected Papers in Biochemistry*, vol. 7, ed. Y. Kaziro [Tokyo: University of Tokyo Press, 1971]).

7. M. B. Hoagland, P. C. Zamecnik, N. Sharon, F. Lipmann, M. P. Stulberg, and P. D. Boyer, "Oxygen Transfer to AMP in the Enzymatic Synthesis of the Hydroxamate of Tryptophan," *Biochimica et Biophysica Acta* 26 (1957): 215–17.

6: Messenger RNA

1. Quoted in Horace Judson, *The Eighth Day of Creation* (New York: Simon & Schuster, 1979), 109–10.

2. Francis Crick, *What Mad Pursuit—A Personal View of Scientific Discovery* (New York: Basic Books, 1988).

3. Paul Zamecnik, "An Historical Acccount of Protein Synthesis, with Current Overtones—A Personalized View," *Cold Spring Harbor Symposia on Quantitative Biology* 34 (1969): 4–5.

4. Arthur B. Pardee, François Jacob, and Jacques Monod, "The Genetic Control and Cytoplasmic Expression of 'Inducibility' in the Synthesis of β-Galactosidase by *E. coli*," *Journal of Molecular Biology* 1 (1959): 165–78.

5. Jacques Monod, *Chance and Necessity* (New York: Alfred A. Knopf, 1971); Judson, *Eighth Day;* André Lwoff and Agnes Ullmann, eds., *Origins of Molecular Biology: A Tribute to Jacques Monod* (New York: Academic Press, 1979).

6. François Jacob, *The Statue Within: An Autobiography* (New York: Basic Books, 1988), 290–321.

7. Quoted in Judson, *Eighth Day*, 433.

8. Monica Riley, Arthur Pardee, François Jacob, and Jacques Monod, "On the Expression of a Structural Gene," *Journal of Molecular Biology* 2 (1960): 216–25.

9. Sydney Brenner, François Jacob, and Matthew Meselson, "An Unstable Intermediate Carrying Information from Genes to Ribosomes for Protein Synthesis," *Nature* 190 (1961): 576–81.

10. F. Gros, H. Hiatt, W. Gilbert, C. G. Kurland, R. W. Risebrough, and J. D. Watson, "Unstable Ribonucleic Acid Revealed by Pulse Labeling of *E. Coli*," *Nature* 190 (1961): 581–85.

11. Marshall W. Nirenberg and J. H. Matthaei, "The Dependence of Cell-Free Protein Synthesis in *E. Coli* upon Naturally Occurring or Synthetic Polyribonucleotides," *Proceedings of the National Academy of Sciences* 47 (1961): 1588–602.

12. Marvin Lamborg and Paul C. Zamecnik, "Amino Acid Incorporation by Extracts of *E. Coli*," *Biochimica et Biophysica Acta* 42 (1960) 206–11.

*7: **Entr'acte***

1. M. B. Hoagland and L. T. Comly, "Interaction of Soluble Ribonucleic Acid and Microsomes," *Proceedings of the National Academy of Sciences,* 46 (1960): 1554–63.

2. M. B. Hoagland, "The Relation of Nucleic Acid and Protein Synthesis as Revealed by Studies in Cell-Free Systems," in E. Chargaff and J. N. Davidson, eds., *The Nucleic Acids* (New York: Academic Press, 1960), chap. 37.

3. M. B. Hoagland, "Nucleic Acids and Proteins," *Scientific American,* Dec. 1959, 55–61.

4. M. B. Hoagland, "Coding, Information Transfer and Protein Synthesis," in J. B. Stanbury, J. B. Wyngaarden, and D. S. Fredrickson, eds., *The Metabolic Basis of Inherited Disease* (New York: McGraw-Hill, 1978; earlier editions in 1966 and 1971).

5. Bernard D. Davis, *Storm over Biology: Essays on Science, Sentiment, and Public Policy* (Buffalo: Prometheus Books, 1986).

6. M. B. Hoagland, O. A. Scornik, and L. C. Pfefferkorn, "Aspects of the Control of Protein Synthesis in Normal and Regenerating Rat Liver; II. A Microsomal Inhibitor of Amino Acid Incorporation Whose Action Is Antagonized by Guanosine Triphosphate," *Proceedings of the National Academy of Sciences* 51 (1964): 1184–91.

7. O. A. Scornik, M. B. Hoagland, L. C. Pfefferkorn, and E. A. Bishop, "Inhibition of Protein Synthesis in Rat Liver Microsome Fractions," *Journal of Biological Chemistry* 242 (1967): 131–39.

8. R. D. Nolan and M. B. Hoagland, "Cytoplasmic Control of Protein Synthesis in Rat Liver," *Biochimica et Biophysica Acta* 247 (1971): 609–20; H. T. R. Rupnial and R. V. Quincey, "Mechanisms of Action of a Microsomal Inhibitor of Protein Synthesis Potentiated by Oxidized Glutathione," *Biochemical Journal* 136 (1973): 335–42.

9. Oscar A. Scornik and Violeta Botbol, "Protein Metabolism and Liver Growth," in *Lysosomes: Their Role in Protein Breakdown* (London: Academic Press, 1987), 445–84.

10. M. B. Hoagland And B. A. Askonas, "Aspects of Control of Protein Synthesis in Normal and Regenerating Rat Liver: I. A. Cytoplasmic RNA-Containing Fraction That Stimulates Amino Acid Incorporation," *Proceedings of the National Academy of Sciences* 49 (1963): 130–37; J. A. A. Gardner and M. B. Hoagland, "The Isolation of Guanosine Tetraphosphate from Commercially Available Preparations of Guanosine Triphosphate," *Journal of Biological Chemistry* 240 (1965): 1244–46; M. H. Dresden and M. B. Hoagland, "Polyribosomes from *E. coli;* Enzymatic Method for Isolation," *Science* 149 (1965): 647–49; idem, "Polyribosomes of *E. coli:* Breakdown during Glucose Starvation," *Journal of Biological Chemistry* 242 (1967): 1065–68; idem, "Polyribosomes of *E. coli:* Reformation during Recovery from Glucose Starvation," ibid., 1069–73; S. H. Wilson and M. B. Hoagland, "Studies on the Physiology of Rat Liver Polyribosomes: Quantitation and Intracellular Distribution of Ribosomes," *Proceedings of the National Academy of Sciences* 54 (1965): 600–607; idem, "Physiology of Rat Liver Polysomes: II. The Stability of Messenger Ribonucleic Acid and Ribosomes," *Biochemical Journal* 103 (1967): 556–66; idem, "Physiology of Rat Liver Polysomes: III. Protein Synthesis by Stable Polysomes," ibid., 567–72; H. Sox and M. B. Hoagland, "Functional Alteration in Rat Liver Polyribosomes during Starvation and Refeeding, *Journal of Molecular Biology* 20 (1966): 113–21; A. I. Meisler and B. E. Troop, "Studies on Ribonucleic Acid Synthesis in Nuclei Isolated from Rat Liver," *Biochimica et Biophysica Acta* 174 (1969): 476–90; S. H. Wilson and R. V. Quincey, "Quantitative Determination of Low Molecular Weight Ribonucleic Acids in Rat Liver Microsomes," *Journal of Biological Chemistry* 244 (1969): 1092–96; R. V. Quincey and S. H. Wilson, "The Utilization of Genes for Ribosomal RNA, 5S RNA and Transfer RNA in Liver Cells of Adult Rats," *Proceedings of the National Academy of Sciences* 64 (1969): 981–88; G. G. Cornwell III and M. B. Hoagland, "The Simultaneous Determination of Transfer Ribonucleic Acid and Amino Acid Attachment to Rat Liver Ribosomes in the Presence of Unfractionated Supernatant," *Biochimica et Biophysica Acta* 238 (1971): 259–63; M. Takagi and M. B. Hoagland, "Polysome-Binding Capacity of Membranes in the Cytoplasmic Extract Prepared from Phenobarbital-Treated and Regenerating Rat Liver," *Biochemical and Biophysical Research Communications* 58 (1974): 868–75; idem, "A Polysome-Membrane Binding System from Rat Liver: I. Basic Characterization of the Binding System," *Journal of Biochemistry* 78 (1975): [illegible]–33.

11. "Worcester Foundation Marks Time," *Nature* 222 (1969): 216.

8: Two Traditions

1. I am indebted to Philip J. Pauly, *Controlling Life: Jacques Loeb and the Engineering Ideal in Biology* (New York: Oxford University Press, 1987), for some of the foregoing analysis of Loeb's influence on the Worcester Foundation's founders.

9: Rejuvenation

1. A. J. Aspen and M. B. Hoagland, "Uncoupling of Amino Acid Turnover on Transfer RNA from Protein Synthesis in HeLa Cells," *Biochimica et Biophysica Acta* 518 (1978): 482–96.

2. Mahlon Hoagland, *The Roots of Life: A Layman's Guide to Genes, Evolution, and the Ways of Cells* (Boston: Houghton Mifflin, 1978; New York: Avon, 1979); idem, *Discovery: The Search for DNA's Secrets* (Boston: Houghton Mifflin, 1981; New York: Van Nostrand Reinhold, 1983).

10: And Some Politics

1. Selma Mushkin, *Biomedical Research: Costs and Benefits* (Cambridge, Mass.: Ballinger, 1979).

2. *Biomedical Discoveries Adopted by Industry for Purposes Other Than Health Services*, NIH report, March 1981.

3. Lewis Thomas, "Biomedical Science and Human Health: The Long-range Prospect," *Daedalus* 106, no. 3 (Summer 1977): 163–71.

Index

Page numbers in *italics* refer to illustrations.